Selected Titles in This Serie

Volume

2 **Gregory F. Lawler and Lester N. Coyle**
Lectures on contemporary probability
1999

1 **Charles Radin**
Miles of tiles
1999

STUDENT MATHEMATICAL LIBRARY
IAS/PARK CITY MATHEMATICAL SUBSERIES
Volume 2

Lectures on Contemporary Probability

Gregory F. Lawler
Lester N. Coyle

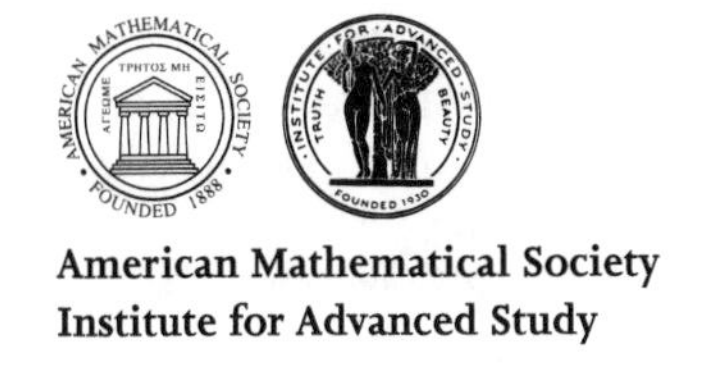

American Mathematical Society
Institute for Advanced Study

1991 *Mathematics Subject Classification.* Primary 60–02;
Secondary 60J10, 60J15, 65C05.

Research of Gregory Lawler partially supported
by the National Science Foundation

ABSTRACT. Lectures given to undergraduates at the Institute for Advanced Study/Park City Institute on Probability in 1996. Topics include: random walk (simple and self-avoiding), Brownian motion, card shuffling, Markov chains and Markov chain Monte Carlo, spanning trees, and simulation of random walks and models in finance. This book could be used as a supplement to or for a second course after a calculus based probability course.

Library of Congress Cataloging-in-Publication Data

Lawler, Gregory F., 1955–
Lectures on contemporary probability / Gregory F. Lawler, Lester N. Coyle.
p. cm. — (Student mathematical library, ISSN 1520-9121 ; v. 2)
Includes bibliographical references.
ISBN 0-8218-2029-X (softcover : alk. paper)
1. Stochastic processes. 2. Probabilities. I. Coyle, Lester N. (Lester Noel), 1967– . II. Title. III. Series.
QA274.L384 1999
519.2–dc21 99-23838
CIP

∞ The paper used in this book is acid-free and falls within the guidelines
established to ensure permanence and durability.
Visit the AMS home page at URL: http://www.ams.org/

10 9 8 7 6 5 4 3 2 1   04 03 02 01 00 99

Contents

IAS/Park City Mathematics Institute

The IAS/Park City Mathematics Institute (PCMI) was founded in 1991 as part of the "Regional Geometry Institute" initiative of the National Science Foundation. In mid 1993 the program found an institutional home at the Institute for Advanced Study (IAS) in Princeton, New Jersey. The PCMI will continue to hold summer programs alternately in Park City and in Princeton.

The IAS/Park City Mathematics Institute encourages both research and education in mathematics, and fosters interaction between the two. The three-week summer institute offers programs for researchers and postdoctoral scholars, graduate students, undergraduate students, high school teachers, mathematics education researchers, and undergraduate faculty. One of PCMI's main goals is to make all of the participants aware of the total spectrum of activities that occur in mathematics education and research: we wish to involve professional mathematicians in education and to bring modern concepts in mathematics to the attention of educators. To that end the summer institute features general sessions designed to encourage interaction among the various groups. In-year activities at sites around the country form an integral part of the High School Teacher Program.

Each summer a different topic is chosen as the focus of the Research Program and Graduate Summer School. Activities in the Undergraduate Program deal with this topic as well. Lecture notes from the Graduate Summer School are published each year in the IAS/Park City Mathematics Series. Course materials from the Undergraduate Program, such as the current volume, are now being published as part of the IAS/Park City Mathematical Subseries in the Student Mathematical Library. We are happy to make available more of the excellent resources which have been developed as part of the PCMI.

At the summer institute late afternoons are devoted to seminars of common interest to all participants. Many deal with current issues in education; others treat mathematical topics at a level which encourages broad participation. The PCMI has also spawned interactions between universities and high schools at a local level. We hope to share these activities with a wider audience in future volumes.

Robert Bryant and Dan Freed, Series Editors

May, 1999

Preface

These notes summarize two of the three classes held for undergraduates at the 1996 Park City/IAS Institute in Probability. There were twenty undergraduates participating, who were divided into an advanced group and a beginning group. Both groups participated in a class on computer simulations in probability. The beginner's class, taught by Emily Puckette, discussed Markov chains and random walks, covering material in the first few chapters of [**L3**]. This book gives notes from the advanced class and the computer class. The first ten lectures are those given to the advanced class by Greg Lawler, and the last three summarize the material in the computer class led by Lester Coyle. The material was coordinated so that some of the major simulations done in the computer class related to topics discussed in the advanced class. For this reason we have decided to combine these notes into one book.

The title of the advanced lecture series (and of these notes) is taken from a recent book [**S**], *Topics in Contemporary Probability and its Applications,* edited by J. Laurie Snell, which contains a number of survey articles that are accessible to advanced undergraduates and beginning graduate students. The lectures were based loosely on three of the papers in that book: "Random Walks: Simple and Self-Avoiding" by Greg Lawler, "How Many Times Should You Shuffle a Deck of Cards?" by Brad Mann, and "Uniform Spanning Trees" by

Robin Pemantle. The idea was to present some topics which are accessible to advanced undergraduates yet are areas of current research in probability.

The first lecture discusses simple random walk in one dimension and is anything but contemporary. It leads to a derivation of Stirling's formula. The second lecture discusses random walk in several dimensions and introduces the notion of power laws. Standard results about probability of return to the origin and the intersection exponent are discussed. The latter is a simply stated exponent whose value is not known rigorously today, and it is a natural exponent to study by simulation. The third lecture discusses the self-avoiding walk, which is a very good example of a simply stated mathematical problem for which most of the interesting questions are still open problems. The fourth lecture considers the continuous limit of random walk, Brownian motion. This topic was included to help those students who were involved in simulations related to finance.

The next two lectures consider the problem of shuffling a deck of cards. Lecture 5 discusses the general idea of random permutations and introduces the notion of a random walk on a symmetric group. The case of random riffle shuffles and the time (number of shuffles) needed to get close to the uniform distribution was analyzed in a paper of Bayer and Diaconis [**BD**], and Mann's paper [**M**] is an exposition of this result. We give a short discussion of this result, although we do not give all the details of the proof. This topic leads naturally to the discussion of Markov chains and rates of convergence to equilibrium. Lecture 7 is a standard introduction to Markov chains; it outlines a proof of convergence to equilibrium that emphasizes the importance of the size of the second eigenvalue for understanding the rate of convergence. Lecture 8 disucsses a recent important technique to sample from complicated distributions, Markov Chain Monte Carlo.

Lecture 9 discusses a very beautiful relationship between random walks and electrical networks (see [**DS**] for a nice exposition of this area). The basic ideas in this section are used in more sophisticated probability; this is basically the discrete version of Dirichlet forms. The work on electrical networks leads to the final lecture on uniform spanning trees. We discuss one result that relates three initially quite different objects: uniform spanning trees, random walks on graphs, and electrical networks.

The purpose of the computer class was to introduce students to the idea of Monte Carlo simulations and to give them a chance to do some nontrivial projects. The previous computer experience of the students varied widely, some having significant programming backgrounds and some having never computed. We first used Maple and then C as the languages for simulations. While these sections are labeled as "lectures" they actually represent a summary of many lectures, and the topics were not really presented in the order that they appear here. Lecture 11 discusses simulations for random walks and includes some basic material on curve fitting to estimate exponents. It ends with a discussion of the most serious project done in this area, the estimate of the intersection exponent. Lecture 12 discusses simulation topics other than random walk that were discussed in the class, including sampling from continuous distributions, random permutations, and finally a more difficult project — using Markov Chain Monte Carlo as discussed in Lecture 8 to estimate the number of matrices with certain conditions. The last lecture discusses a different area, simulations of stochastic differential equations for applications in finance.

We conclude the book with a number of problems that were presented to the students. The difficulty of these problems varies greatly; some are routine, but many were given more to stimulate thought than with the expectation that the students would completely solve them. They are numbered to indicate which lecture they refer to. Of particular note are the problems from Lectures 11 and 12. These are representative of the simpler projects that we gave to the students as they were learning how to do simulations, and are typical of simulation problems that we give to students when we teach undergraduate probability.

We would like to thank a number of people who helped with the program for undergraduates, including: Emily Puckette, the third member of our team; Chad Fargason, who helped write some of the software used in the labs; David Levin, who helped in the preparation of these notes; Brad Mann and Robin Pemantle, for providing copies of their papers; Persi Diaconis, who suggested one of the ideas that is used in Lecture 8; Jennifer Chayes, Thad Dankel, Ingemar Kaj, Ariel Landau, and Vladimir Vinogradov, who gave special lectures to the undergraduates; and most of all to the students who participated.

The first author was supported in part by the National Science Foundation.

Lecture 1

Simple Random Walk and Stirling's Formula

We will start our discussion of contemporary probability by discussing an old problem — at least a few hundred years old. Imagine a walker moving randomly on the integers. The walker starts at 0 and at every integer time n the walker flips a fair coin and moves one step to the right if it comes up heads and one step to the left if it comes up tails. The position of the walker at time n will be denoted by S_n. S_n is a random variable; the position depends on the outcomes of the n flips of the coin. We assume $S_0 = 0$, and we can write

$$S_n = X_1 + \cdots + X_n,$$

where $X_i = 1$ if the ith flip was heads and $X_i = -1$ if the ith flip was tails. We assume that the coin flips are independent, so that the random variables $X_1, X_2, \ldots$ are indepedent. Three natural questions to ask are:

— About how far does the walker go in n steps?

— Where do we expect the walker to be after n steps? (The answer to this should involve a probability distribution for the possible positions.)

— Does the walker always return to the starting point; or more generally, is every integer visited infinitiely often by the walker?

We can give a partial answer to the first question fairly easily. It is clear that the expected distance $\mathbf{E}(S_n)$ is zero because the probability of being at j is the same as being at $-j$. Computing $\mathbf{E}(|S_n|)$ is not so easy; however, it is straightforward to calculate $\mathbf{E}(S_n^2)$. Note that

$$\begin{aligned}\mathbf{E}(S_n^2) &= \mathbf{E}[(X_1 + \cdots + X_n)^2] \\ &= \sum_{j=1}^{n} \mathbf{E}(X_j^2) + \sum_{i \neq j} \mathbf{E}(X_i X_j).\end{aligned}$$

Since $X_j^2 = 1$ and $\mathbf{E}(X_i X_j) = \mathbf{E}(X_i)\mathbf{E}(X_j) = 0$ if $i \neq j$, we see that

$$\mathbf{E}(S_n^2) = n.$$

This says that the expected *squared* distance is n, and we can infer (at least informally) that the expected distance should be of order $\sqrt{n}$ (it is actually about $c\sqrt{n}$ for a constant c different than 1).

Where do we expect to be after n steps? Since the walker always moves from an even integer to an odd integer or from an odd integer to an even integer, we know for sure that we are at an even integer if n is even or an odd integer if n is odd. Let us suppose that we have gone $2n$ steps, and ask what is the probability that we are at the even integer $2j$. Using the binomial distribution, we can determine the probability exactly. There are 2^{2n} different sequences of $+1$'s and -1's of length $2n$. Each one has probability 2^{-2n} of occurring. In order for S_{2n} to equal $2j$ we need $n + j$ moves to the right $(+1)$ and $n - j$ moves to the left (-1). The number of sequences with $n + j$ "$+1$"s and $n - j$ "-1"s is given by the appropriate binomial coefficient. Hence

$$\mathbf{P}\{S_{2n} = 2j\} = \binom{2n}{n+j} 2^{-2n} = 2^{-2n} \frac{(2n)!}{(n+j)!(n-j)!}.$$

In particular,

$$\mathbf{P}\{S_{2n} = 0\} = 2^{-2n} \frac{(2n)!}{n!n!}.$$

This is a nice exact expression. Unfortunately, it is not so easy to see how big or small it is. What we need is an approximate formula for $n!$. There is such a formula, first published by J. Stirling in 1730 and now known as Stirling's formula. It is our goal here to see how this formula can be derived.

How do we give a good approximation of $n!$ for large n? Note that

$$n! = 1 \cdot 2 \cdots (n-1) \cdot n < n^n.$$

But this is not a very good bound. One might try averaging the terms between $1, \ldots, n$ and estimating $n!$ by $(n/2)^n = 2^{-n} n^n$. However, a little computation (with a computer, say) would show that this is not very good. This leads, however, to trying $a^{-n} n^n$ for some number $a > 1$. What a should we choose? Let's take logarithms. Clearly,

$$\ln[a^{-n} n^n] = n(\ln n - \ln a).$$

To estimate $\ln n! = \ln 1 + \ln 2 + \cdots + \ln n$ we approximate a sum by an integral:

$$\int_1^n \ln(x)\, dx \leq \sum_{j=1}^{n} \ln j \leq \int_1^{n+1} \ln(x)\, dx.$$

By doing the integral (a standard exercise in integration by parts) we get

$$\begin{aligned} n(\ln n - 1) + 1 &\leq \ln n! \\ &\leq (n+1)\ln(n+1) - n \\ &= n(\ln n - 1) + \ln(n) + (n+1)\ln\left[1 + \frac{1}{n}\right]. \end{aligned}$$

The error term $\epsilon_n = \ln n + (n+1)\ln[1 + (1/n)]$ satisfies

$$\frac{\epsilon_n}{n} \to 0, \quad n \to \infty$$

(why?), so our best guess for a satifies $\ln a = 1$, i.e., $a = e$. We now have the approximation

$$n! \approx n^n e^{-n}.$$

It is a good approximation in the sense that

$$\lim_{n\to\infty} \frac{\ln n!}{\ln[n^n e^{-n}]} = 1.$$

We now ask the question: does the limit

$$\lim_{n\to\infty} \frac{n!}{n^n e^{-n}}$$

exist? Let $x_n = n![n^n e^{-n}]^{-1}$ and $b_n = x_{n-1}/x_n$ (where we set $x_0 = 1$ for convenience). Then

$$\begin{aligned} x_n^{-1} &= b_1 \cdots b_n \\ &= x_1^{-1} \prod_{j=2}^{n} e^{-1} \left[1 + \frac{1}{j-1}\right]^{j-1} = e^{-1} \prod_{j=1}^{n-1} e^{-1} \left[1 + \frac{1}{j}\right]^{j}. \end{aligned}$$

We all know from calculus that

$$\lim_{n\to\infty} \left(1 + \frac{1}{n}\right)^n = e.$$

We therefore write

$$x_{n+1}^{-1} = e^{-1} \prod_{j=1}^{n} [1 + \epsilon_j],$$

where

$$\epsilon_j = e^{-1} \left[1 + \frac{1}{j}\right]^j - 1 \to 0.$$

Suppose we have numbers $a_j \to 0$. When can we conclude that the limit

$$\lim_{n\to\infty} \prod_{j=1}^{n} [1 + a_j]$$

exists? (Assume, to avoid trivial cases, that $a_j \neq -1$ for all j.) A necessary condition is that $a_j \to 0$. But this is not sufficient. Let us assume that $a_j \to 0$. When dealing with large products it is often easier to take logarithms, since logarithms convert products to sums. We know from properties of limits that

$$\ln \lim_{n\to\infty} \prod_{j=1}^{n} [1 + a_j] = \lim_{n\to\infty} \ln \prod_{j=1}^{n} [1 + a_j],$$

with each limit existing if and only if the other limit exists. Also,

$$\ln \prod_{j=1}^{n} [1 + a_j] = \sum_{j=1}^{n} \ln[1 + a_j].$$

We need to approximate $\ln(1+y)$ for small y — Taylor polynomials are the necessary tool. Recall that

$$\ln(1+x) = x - \frac{1}{2}x^2 + R_3(x),$$

where $R_3(x)$ is a remainder term satisfying

$$|R_3(x)| = C|x|^3, \quad |x| < 1/2.$$

(You may wish to review Taylor's theorem with remainder to find such a C.) Mathematicians often write this as

$$\ln(1+x) = x - \frac{1}{2}x^2 + O(x^3),$$

where $O(x^3)$ represents a function that is bounded by a constant times x^3. In particular, for all x small enough

$$|\ln(1+x)| \leq 2|x|.$$

In particular, *if the $\{a_j\}$ is an absolutely convergent series, then the limit*

$$\lim_{n\to\infty} \ln \prod_{j=1}^{n} [1+a_j]$$

exists.

Returning to the problem at hand, let

$$\epsilon_j = e^{-1}\left[1+\frac{1}{j}\right]^j.$$

To see how close this is to 1, we again take logs and expand in the Taylor series,

$$\ln\left[1+\frac{1}{j}\right]^j = j\ln\left[1+\frac{1}{j}\right] = j\left[\frac{1}{j} - \frac{1}{2j^2} + O\left(\frac{1}{j^3}\right)\right]$$
$$= 1 - \frac{1}{2j} + O\left(\frac{1}{j^2}\right).$$

Using the expansion for the exponential,

$$e^x = 1 + x + O(x^2), \quad x \to 0$$

we get

$$\left[1+\frac{1}{j}\right]^j = \exp\left\{1 - \frac{1}{2j} + O\left(\frac{1}{j^2}\right)\right\} = e\left[1 - \frac{1}{2j} + O\left(\frac{1}{j^2}\right)\right],$$

or

$$e^{-1}\left[1+\frac{1}{j}\right]^j - 1 = -\frac{1}{2j} + O\left(\frac{1}{j^2}\right).$$

So, for j large, ϵ_j is about $-1/2j$. This is unfortunate since

$$\sum_{j=1}^{\infty} \frac{1}{2j} = \infty.$$

From this one can show that the limit $\lim x_n$ as defined above does NOT exist.

Let us try to improve the approximation, by including a power of n, i.e., we will approximate $n!$ by $n^n e^{-n} n^s$, where s is some number independent of n. What value of s shall we choose? Let $y_n = n![n^n e^{-n} n^s]^{-1}$, $c_0 = 1$ and $c_n = y_{n-1}/y_n$, so that

$$y_{n+1}^{-1} = c_1 \cdots c_{n+1} = y_1^{-1} \prod_{j=2}^{n} b_j \left[\left(\frac{j+1}{j}\right)\right]^s.$$

We have already seen that

$$b_j = 1 - \frac{1}{2j} + O\left(\frac{1}{j^2}\right).$$

For a fixed s, we can do another Taylor expansion and see that

$$\left(\frac{j+1}{j}\right)^s = \left(1+\frac{1}{j}\right)^s = 1 + s\frac{1}{j} + O\left(\frac{1}{j^2}\right).$$

We now choose $s = 1/2$ so that the $(1/j)$ terms will cancel. If $s = 1/2$,

$$b_j \left(1+\frac{1}{j}\right)^s = 1 + O\left(\frac{1}{j^2}\right).$$

But, $\sum j^{-2} < \infty$, so we can conclude that the limit

$$\lim_{n\to\infty} y_n^{-1} = \lim_{n\to\infty} \prod_{j=1}^{n} c_j = e^{-1} \lim_{n\to\infty} \prod_{j=1}^{n} \left[1 + O\left(\frac{1}{j^2}\right)\right]$$

exists. We finally have shown the following: *There exists a positive number L such that*

$$\lim_{n\to\infty} \frac{n!}{n^n e^{-n}\sqrt{n}} = L.$$

This method does not determine the value L. We will determine it in another way in the problems (see Problems 1-3 to 1-6). Returning to the original problem about random walks, we see that

$$\mathbf{P}\{S_{2n}=0\}=2^{-2n}\frac{2n!}{n!n!}\sim 2^{-2n}\frac{L(2n)^{2n}e^{-2n}\sqrt{2n}}{[Ln^{n}e^{-n}\sqrt{n}]^{2}}=\frac{\sqrt{2}}{L\sqrt{n}}.$$

So the probability of being at the origin is about a constant times $n^{-1/2}$. This is consistent with what we already know. We have seen that the random walker tends to go a distance about a constant times $\sqrt{n}$. There are $c\sqrt{n}$ such integer points, so it is very reasonable that a particular one is chosen with probability a constant times $n^{-1/2}$.

We now consider the total number of times that the walker visits the origin. Let R_n be the number of visits to the origin up through time $2n$. Then

$$R_n = Y_0 + Y_1 + \cdots + Y_n,$$

where $Y_j = 1$ if $S_{2j} = 0$ and $Y_j = 0$ if $S_{2j} \neq 0$. Note that $\mathbf{E}(Y_j) = \mathbf{P}\{S_{2j} = 0\}$. Therefore,

$$\mathbf{E}(R_n) = \mathbf{E}(Y_0) + \cdots + \mathbf{E}(Y_n) = \sum_{j=0}^{n} \mathbf{P}\{S_{2j} = 0\}$$

$$\sim \sum_{j=1}^{n} \frac{\sqrt{2}}{L} j^{-1/2} \sim \frac{2\sqrt{2}n^{1/2}}{L}.$$

(Why is the last step true?) In particular, the expected number of visits goes to infinity as $n \to \infty$. This indicates (and we will discuss how to verify it in the next lecture) that the number of visits is infinite. (There is a subtle point here. If

$$R = R_\infty = \sum_{j=1}^{\infty} Y_j$$

is the total number of visits to the origin, then we have demonstrated that $\mathbf{E}(R) = \infty$. What we would like to conclude is that $R = \infty$. There are positive finite random variables X with $\mathbf{E}(X) = \infty$ (see Problem 1-10), so in order to show that $R = \infty$ we need to show more than $\mathbf{E}(R) = \infty$.)

Lecture 2

Simple Random Walk in Many Dimensions

We will now allow our random walker to move in two, three, or even d dimensions. We will use $\mathbf{Z}^d$ to denote the d-dimensional integer grid, i.e., the set of d-tuples $x = (x^1, \ldots, x^d)$ where the x^i are all integers. Each x in $\mathbf{Z}^d$ has $2d$ "nearest neighbors", points that are at a distance one away from it. To do simple random walk on $\mathbf{Z}^d$, the walker starts at the origin and at each integer time n moves to one of the nearest neighbors, each with probability $1/2d$. We let $S_n = (S_n^1, \ldots, S_n^d)$ be the position of the walker after n steps. Then

$$S_n = X_1 + \cdots + X_n,$$

where $X_i = (X_i^1, \ldots, X_i^d)$ and the X_i are independent random "vectors" with

$$\mathbf{P}\{X_i = y\} = 1/2d,$$

for each $y \in \mathbf{Z}^d$ that is distance one from the origin. Equivalently, we can say that there are $(2d)^n$ "random walk paths" of n steps that start at the origin, and each one of these paths gets the same probability, $(2d)^{-n}$. We will ask the same questions for this multidimensional walk that we asked for the one-dimensional walk.

Let us start by asking, how far the random walker goes. Again, it is easier to compute the expected *squared* distance from the origin,

$$\begin{aligned}\mathbf{E}[|S_n|^2] &= \mathbf{E}[(S_n^1)^2 + \cdots + (S_n^d)^2] \\ &= \mathbf{E}[(S_n^1)^2] + \cdots + \mathbf{E}[(S_n^d)^2] = d\mathbf{E}[(S_n^1)^2].\end{aligned}$$

Similarly as before, we calculate

$$\mathbf{E}[(S_n^1)^2] = \mathbf{E}[(X_1^1 + \cdots + X_n^1)^2] = \sum_{j=1}^{n} \mathbf{E}[(X_j^1)^2] + \sum_{i \neq j} \mathbf{E}[X_i^1 X_j^1].$$

As before, for $i \neq j$, $\mathbf{E}[X_i^1 X_j^1] = \mathbf{E}[X_i^1]\mathbf{E}[X_j^1] = 0$. Since the probability that the walker moves in the first coordinate (either $+1$ or -1) is $1/d$, $\mathbf{E}[(X_j^1)^2] = 1/d$. From this we see that

$$\mathbf{E}[|S_n|^2] = n,$$

regardless of the dimension.

What is the probability that the random walker is at the origin 0 after n steps? Note that the random walker always moves from even points to odd points or from odd points to even points, where in this case we call a point $x = (x^1, \ldots, x^d)$ even if the sum of the components is even. So let us consider an even time $2n$. If n is large, then the law of large numbers tells us that approximately $(2n/d)$ of the steps will be in each of the d component directions. In order for the walker to be at the origin after $2n$ steps, the walker will have had to have taken an even number of steps in each component. The probability of doing this will be about $2^{(1-d)}$ (why is this correct?). Assuming that an even number of steps are taken in each component, then we also need that the ith component of the walker is 0. The probability of this happening in the ith component is about the same as the probability that a one-dimensional walker is at the origin after $(2n/d)$ steps. From the previous lecture (using Stirling's formula $n! \sim \sqrt{2\pi} n^n e^{-n} \sqrt{n})$), this probability is approximately $[\sqrt{\pi(n/d)}]^{-1}$. Since there are d components, all of which have to be zero at time $2n$, we get the approximation

$$\mathbf{P}\{S(2n) = 0\} \sim 2^{1-d} \left(\frac{1}{\sqrt{\pi(n/d)}} \right)^d = \left(\frac{d^{d/2}}{2^{d-1}\pi^{d/2}} \right) n^{-d/2}.$$

The fact that the probability is about a constant times $n^{-d/2}$ makes sense: the walker tends to be distance about $\sqrt{n}$ from the origin, and there are about $n^{d/2}$ points in $\mathbf{Z}^d$ that are within distance $\sqrt{n}$ from the origin. Hence, we would expect that the probability of choosing a particular one would be of order $n^{-d/2}$.

How often does the random walker visit the origin? As before, let $Y_j = 1$ if $S_{2j} = 0$ and otherwise $Y_j = 0$. Then the number of visits up through time $2n$ is given by the random variable

$$R_n = Y_0 + \cdots + Y_n,$$

and the total number of visits is given by

$$R = R_\infty = Y_0 + Y_1 + \cdots .$$

Note that

$$\mathbf{E}[R_n] = \sum_{j=0}^{n} \mathbf{P}\{S_{2j} = 0\}.$$

We now see a big difference depending on the dimension. Suppose d is at least three. Then

$$\mathbf{E}[R] = \sum_{j=0}^{\infty} \mathbf{P}\{S_{2j} = 0\} \leq 1 + \text{const} \sum_{j=1}^{\infty} j^{-d/2} < \infty.$$

The expected number of returns is finite. In fact, we can ask if the random walker ever returns to the origin, i.e., if $R > 1$. Suppose the walker returns at some time $2j > 0$. Then the expected number of returns from then on is the same as for the walker starting at the origin. From this we get the relation

$$\mathbf{E}[R \mid R > 1] = 1 + \mathbf{E}[R].$$

But,

$$\begin{aligned} \mathbf{E}[R] &= 1 \cdot \mathbf{P}\{R = 1\} + \mathbf{P}\{R > 1\}\mathbf{E}[R \mid R > 1] \\ &= 1 + \mathbf{P}\{R > 1\}\mathbf{E}[R], \end{aligned}$$

or

$$\mathbf{P}\{R = 1\} = \frac{1}{\mathbf{E}[R]} > 0.$$

Hence in three or more dimensions the probability of returning to the origin is strictly less than 1.

For $d = 2$,

$$\mathbf{E}[R_n] = \sum_{j=0}^{n} \mathbf{P}\{S_{2j} = 0\} \sim 1 + \sum_{j=1}^{n} \frac{1}{\pi j} \sim \frac{1}{\pi} \ln n$$

(why is the last step valid?). In particular, $\mathbf{E}[R] = \infty$; hence the expected number of returns is infinite. Let us consider q_n, the probability that the walker goes $2n$ steps without returning to the origin,

$$q_n = \mathbf{P}\{R_n = 1\} = \mathbf{P}\{S_2 \neq 0, S_4 \neq 0, \cdots, S_{2n} \neq 0\}.$$

Suppose $R_n > 1$ and consider

$$\mathbf{E}[R_n \mid R_n > 1].$$

Take the first positive time, say $2j > 0$, that the walker is at the origin. Assume $j \leq n$. By considering the next n steps after time $2j$, we can see that

$$\mathbf{E}[R_n \mid R_n > 1] \leq 1 + \mathbf{E}[R_n]$$

(why is it $\leq$ rather than $=$?). Hence

$$\begin{aligned} \mathbf{E}[R_n] &= 1 \cdot \mathbf{P}\{R_n = 1\} + \mathbf{P}\{R_n > 1\}\mathbf{E}[R_n \mid R_n > 1\} \\ &\leq 1 + \mathbf{P}\{R_n > 1\}\mathbf{E}[R_n], \end{aligned}$$

or

$$q_n = \mathbf{P}\{R_n = 1\} \leq \frac{1}{\mathbf{E}[R_n]} \sim \frac{\pi}{\ln n}.$$

In particular, $q_n \to 0$ as $n \to \infty$. Therefore, if $d = 2$, the random walk is *recurrent*, i.e., always returns to the origin (and, in fact, returns infinitely often). We say that random walk in dimensions three and greater is *transient*.

Now consider $d \geq 3$ and assume we have two random walkers, the positions of which we will denote by S_n and W_n. They both start at the origin. The first walker leaves a red mark at every site visited after time 0 while the second walker leaves a blue mark. The question we ask is: is it guaranteed that some site will have both a red mark and a blue mark? To phrase this more precisely, let p_n be the probability that the points visited up to time n by the first walker and the points visited by the second walker have no intersection,

$$p_n = \mathbf{P}\{S_i \neq W_j, 1 \leq i, j \leq n\}.$$

Then is $p_\infty = \lim_{n\to\infty} p_n > 0$? This is a somewhat tricky question, and we will start by considering the *expected number* of intersections. Let $Y(i,j)$ be the random variable that equals 1 if $S_i = W_j$ and equals 0 otherwise. Let

$$G_n = \sum_{i=1}^{n}\sum_{j=1}^{n} Y(i,j)$$

be the number of pairs (i,j) with $1 \leq i,j \leq n$ and $S_i = W_j$, and let $G = G_\infty$. Then

$$\mathbf{E}[G_n] = \sum_{i=1}^{n}\sum_{j=1}^{n} \mathbf{P}\{S_i = W_j\}.$$

A little bit of thought will convince you that the probability that $S_i = W_j$ is the same as the probability that a walk of $i+j$ steps is at the origin. Hence

$$\mathbf{E}[G_n] = \sum_{i=1}^{n}\sum_{j=1}^{n} \mathbf{P}\{S_{i+j} = 0\}.$$

Recalling that $\mathbf{P}\{S_{2n} = 0\}$ goes to zero like $n^{-d/2}$, we can analyze this sum and show that (up to multiplicative constants) $\mathbf{E}[G_n]$ looks like $n^{1/2}$, if $d = 3$; it looks like $\ln n$ if $d = 4$; and it is bounded by some constant $C < \infty$ (independent of n) if $d \geq 5$. In particular,

$$\mathbf{E}[G] \begin{cases} = \infty, & d = 3, 4, \\ < \infty, & d = 5. \end{cases}$$

This indicates (and it can be proven) that $G = \infty$ if $d = 3, 4$; and for $d \geq 5$ there is a positive probability that $G = 0$,

$$\lim_{n\to\infty} p_n \begin{cases} = 0, & d \leq 4, \\ > 0, & d \geq 5. \end{cases}$$

Four is the critical dimension for intersections of random walks. Another way to think of this fact is to consider the "dimension" of the set of points visited by a random walker. Suppose $d \geq 3$. Then the random walker goes distance approximately $\sqrt{n}$ in time n. Since one tends to visit sites only a finite number of times (by transience of the walk), this means the ball of radius r $(= \sqrt{n})$ contains a constant times r^2 points that are visited by the random walk. This means that the set of points visited is a "two-dimensional" set. When do two-dimensional sets intersect? In less than four dimensions, two-dimensional sets (planes, for example) tend to intersect, while in more than four dimensions they tend not to have an intersection.

Let q_n be the probability that the paths of two random walkers S and W have not intersected by time n,

$$q_n = \mathbf{P}\{S_i \neq W_j, 1 \leq i, j \leq n\}.$$

For $d \leq 4$, $q_n \to 0$. How fast does this probabilty go to zero? The *expected* number of (i, j) with $S_i = W_j$ can be estimated pretty much as above: it looks like a constant (depending on the dimension) times

$$\begin{array}{ll} n^{3/2}, & d = 1, \\ n, & d = 2, \\ n^{1/2}, & d = 3, \\ \ln n, & d = 4. \end{array}$$

When we considered the problem of returns to the origin in one and two dimensions, we found that the probability of no return in n steps looks like the inverse of the expected number of returns. We might hope that this would be the case here. However, it is not. For $d = 1$, it can be shown (using the result from Problems 1-7 to 1-9) that q_n looks like a constant times n^{-1}. For $d = 4$, it can be shown that q_n looks like (const)$(\ln n)^{-1/2}$. For $d = 2, 3$ it is known that q_n decays like $n^{-\zeta}$ for some $\zeta = \zeta(d)$, but the value is not known rigorously! Finding this exponent is an open problem in probability. For $d = 2$, it has been conjectured that $\zeta = 5/8$, while for $d = 3$, numerical simulations indicate a value between .28 and .29. Some rigorous bounds exist:

$$\frac{1}{2} + \frac{1}{8\pi} \leq \zeta < \frac{3}{4}, \quad d = 2,$$

$$\frac{1}{4} \leq \zeta < \frac{1}{2}, \quad d = 3.$$

See Lecture 11 for some approximations of ζ using Monte Carlo simulations.

Lecture 3

Self-Avoiding Walk

Random walks are used in physical chemistry to model polymer chains. Roughly speaking, a polymer chain is a collection of bonds (monomers) attached together randomly with the proviso that the chain cannot cross itself. The study of polymer chains leads to the study of random walks on the integer lattice that are restricted so they do not visit any site more than once. A *self-avoiding walk (SAW)* of length n in the d-dimensional lattice $\mathbf{Z}^d$ is a sequence of points $\omega = (\omega_0, \omega_1, \dots, \omega_n)$ with $\omega_0 = 0$; $|\omega_j - \omega_{j-1}| = 1, j = 1, \dots, n$; and $\omega_i \neq \omega_j$ for $i \neq j$. In other words, a SAW is a simple random walk which does not cross itself. While the SAW seems similar to the simple random walk, it turns out to be much more difficult to analyze. In fact, many of the most interesting questions about SAWs remain open problems today.

Let Ω_n denote the set of self-avoiding walks of length n. The first question we can ask is how many elements does Ω_n have? Let C_n denote the cardinality (number of elements) of Ω_n. The number of simple random walks is $(2d)^n$, so this gives an upper bound for C_n; we can improve this bound slightly by noting that the walk can never return to the site it was most recently visiting, i.e., after the first step there are at most $2d - 1$ nearest neighbors that have not been visited. For a lower bound, we can see that the walk which moves only in the positive direction in each component must be self-avoiding.

The number of walks which move only in the positive direction is d^n. Therefore,

$$d^n \leq C_n \leq 2d(2d-1)^{n-1}. \tag{3.1}$$

This leads us to conjecture that there is some number β with $C_n \approx \beta^n$. Let us see that this is true. Any $(n+m)$-step SAW consists of an n-step SAW joined to an m-step SAW, although not every n-step SAW and m-step SAW can be joined together to form an $(n+m)$-step SAW. We therefore see that

$$C_{n+m} \leq C_n C_m.$$

Let

$$f(n) = \ln C_n.$$

Then f is a *subadditive* function,

$$f(n+m) \leq f(n) + f(m).$$

If f is any subadditive function, then

$$\lim_{n\to\infty} \frac{f(n)}{n} = \inf_n \frac{f(n)}{n}$$

(here inf stands for infimum, which is the same as the greatest lower bound). To see this, let I be the infimum on the right hand side. It suffices to show that for every $\epsilon > 0$ and all n sufficiently large,

$$\frac{f(n)}{n} \leq I + \epsilon.$$

Choose $\epsilon > 0$ and find some m with

$$\frac{f(m)}{m} \leq I + \frac{\epsilon}{2}.$$

For any $n > m$, we can write $n = rm + s$, where r, s are integers and $0 \leq s < m$. Then

$$f(n) = f(rm+s) \leq f(rm) + f(s) \leq rf(m) + f(s).$$

Hence,

$$\frac{f(n)}{n} \leq \frac{rf(m)+f(s)}{rm+s} \leq \frac{f(m)}{m} + \frac{K(m)}{n},$$

where $K(m)$ is the maximum of $f(0), \ldots, f(m-1)$. By choosing n sufficiently large we can guarantee that $K(m)/n \leq \epsilon/2$, and hence that the right hand side is less than $I + \epsilon$.

Applying this result to C_n, we see that the limit

$$\lim_{n \to \infty} \frac{\ln C_n}{n} = a$$

exists, and hence

$$\lim_{n \to \infty} C_n^{1/n} = \beta,$$

for some number $\beta = \beta_d = e^a$. This number is called the connective constant. By (3.1), we can see that $d \leq \beta \leq 2d-1$. The exact value of β is not known; in fact, there is no reason to believe that this number can be written in terms of numbers we already know (π, e, etc.). For $d = 2$, β is between 2.6 and 2.7. For large d, β is just slightly smaller than $2d - 1$.

We can write

$$C_n = \beta^n r(n),$$

where $r(n)^{1/n} \to 1$. What can we say about the correction term $r(n)$? If we were considering simple random walk paths, then the number of walks of length n would be exactly $(2d)^n$, so that there would be no correction term $r(n)$. However, there is another simple random walk quantity that is analogous to the term $r(n)$. Consider C_{2n}. Any $2n$-step self-avoiding walk is obtained by attaching two n-step walks together. Of course, not all combinations of two n-step SAWs produces a $2n$-step SAW. In fact,

$$h(n) = \frac{r(2n)}{r(n)r(n)} = \frac{C_{2n}}{C_n C_n}$$

represents the probability (with respect to the uniform measure on SAWs) that two n-step SAWs of length n have no points of intersection. If we ask for the probability that two simple walks of length n have no point of intersection, then we are considering the quantity q_n discussed in the previous lecture. Recall that

$$q_n \sim \text{const} \begin{cases} n^{-\zeta}, & d = 2, 3, \\ (\ln n)^{-1/2}, & d = 4, \\ 1, & d \geq 5 \end{cases}$$

(where $\zeta = \zeta_d$ depends on the dimension). This leads us to think that

$$h_n \sim \text{const } n^{-\alpha}, \quad d < 4,$$

$$h_n \sim \text{const } (\ln n)^{-\alpha}, \quad d = 4,$$

$$h_n \sim \text{const}, \quad d > 4.$$

Numerical simulations and nonrigorous analysis suggest that $\alpha = 11/32$ in two dimensions; $\alpha \approx .16$ in three dimensions; and $\alpha = 1/4$ in four dimensions. However, there is no proof of these facts; in fact, there is no proof that h_n satisfies this kind of power law. Above four dimensions, recent work of Hara and Slade shows that h_n approaches a nonzero constant. Given the result for the simple random walk, this is not surprising. SAWs should be "thinner" than simple random walks and hence should be less likely to intersect. This is just intuition, and the actual proof is very difficult. The critical dimension for the SAW is four dimensions — the behavior below four dimensions is different from that above four dimensions. The reason that four is the critical dimension is that random walk paths tend to intersect below four dimensions and tend not to intersect above four dimensions. This intuition is nice, but proving things about the SAW is very difficult: above four dimensions it takes the work of Hara and Slade; below four dimensions nobody knows how to prove interesting results.

How far do self-avoiding walks go? Recall that simple random walks of length n tend to be on the order of $\sqrt{n}$ from the origin. This clearly is not true for SAWs in one dimension, where there are only two possible walks. In this case the walk goes distance n in n steps. How about in other dimensions? Above four dimensions we might expect that SAWs look like standard random walks. In fact, Hara and Slade have proved that for $d \geq 5$, the mean squared distance (the expected value of $|\omega_n|^2$) looks like a constant times n, and so the typical SAW is at a distance of order $\sqrt{n}$ from the origin. What about in low dimensions? A chemist, Flory, conjectured that the mean-squared distance should satisfy

$$\mathbf{E}[|\omega_n|^2] \approx n^{6/(d+2)}, \quad d \leq 4.$$

If $d = 1$ this gives n^2, which is clearly correct. Numerical simulation and other nonrigorous techniques indicate that $n^{3/2}$ is correct for $d = 2$, but for $d = 3$ it looks more like $n^{1.18\cdots}$ (so Flory's guess is just a little high in three dimensions). In four dimensions, a logarithmic correction is expected:

$$\mathbf{E}[|\omega_n|^2] \approx n(\ln n)^{1/4}.$$

Proving these facts (especially the $d = 2, 3$ facts) is beyond the techniques known to probabilists today. Even trying to test these conjectures by computer is tricky—it is a nontrivial problem to generate self-avoiding walks on the computer in an efficient way such that all SAW of a certain length are equally likely to be chosen.

We have discussed three numbers associated with the SAW: the connective constant β; the exponent α associated with h_n and $r(n)$; and the exponent associated with the expected mean squared distance of the SAW. The latter two numbers are examples of *critical exponents* and are considered more fundamental than the first number. The reason is that if we make some minor changes to the SAW, such as allowing the walker to take steps either of length one or length two, then the connective constant will change. However, it is conjectured that the critical exponents will not change. The exponents depend only on the dimension of the walk. The independence of the exponents on the fine details of a model is an example of what physicists call *universality*.

Perhaps the most fundamental result of probability theory, the central limit theorem, can be considered as an example of universality. For any independent, identically distributed random variables $X_1, X_2, \ldots$ with mean 0 and finite variance σ^2, the limiting distribution of

$$\frac{X_1 + \cdots + X_n}{\sqrt{n}\sigma}$$

is the standard normal distribution. In other words, fine details (the exact distribution of the X_i) become irrelevant as n gets large.

Lecture 4

Brownian Motion

Brownian motion is a model of continuous random motion. There are a number of ways of defining Brownian motion; we will look at it as a limit of simple random walk as we let the time increment and space increment go to 0. Let S_n denote a simple, one-dimensional random walk as before,

$$S_n = X_1 + \cdots + X_n.$$

This defines S_n for integer n. We define S_t for all real $t \geq 0$ by linear interpolation,

$$S_t = S_n + (t - n)X_{n+1}, \quad n \leq t \leq n + 1.$$

In this way, S_t is a continuous, random function of t.

Simple random walk has time increments $\Delta t = 1$ and space increments $\Delta x = 1$, i.e., the walker moves once every 1 time unit and when it moves it moves a distance of 1 unit. We could instead assume that the walker has some other time increment Δt and some other space increment Δx. Brownian motion is obtained by letting the time and space increment approach 0. We must be careful how we take the limit, however. For ease, let us assume that $\Delta t = 1/N$ for some (large) integer N. We let

$$B_t^{(N)} = (\Delta x) S_{Nt},$$

where Δx is still to be chosen. Brownian motion will be the limit as $N \to \infty$ of the $B_t^{(N)}$. How do we choose the Δx? Our first guess might be to choose $\Delta x = \Delta t = 1/N$. However, the law of large numbers tells us that

$$\lim_{N\to\infty} \frac{X_1 + \cdots + X_N}{N} = 0,$$

and hence with this choice we would have

$$\lim_{N\to\infty} B_1^{(N)} = \lim_{N\to\infty} (1/N) S_N = 0.$$

This would give $B_1 = 0$ (and similarly $B_t = 0$ for all t), which would not be very interesting. We would like to choose Δx so that B_t is a nontrivial random variable. The central limit theorem tells us that the random variables

$$\frac{X_1 + \cdots + X_N}{\sqrt{N}}$$

approach a standard (mean zero, variance 1) normal distribution. For this reason we choose $\Delta x = 1/\sqrt{N} = \sqrt{\Delta t}$.

Brownian motion B_t is the "limit" of the processes $B_t^{(N)}$. We will not discuss the exact nature of the limiting operation, but rather we will try to describe what this process looks like. For each t, there is a random variable B_t which gives the position of a random particle at time t. The particle takes jumps that are independent: more precisely, if $s < t$ then the random variables B_s and $B_t - B_s$ are independent. We have already seen that the central limit theorem tells us that the distribution of B_1 should be a normal distribution with mean zero and variance one. Similarly, we can show that B_t should have a normal distribution with mean 0 and variance t, and more generally, $B_t - B_s$ should have a normal distribution with mean 0 and variance $t - s$. Recall that if X, Y are independent normal random variables with means μ_x, μ_y and variances σ_x^2, σ_y^2 then $X + Y$ has a normal distribution with mean $\mu_x + \mu_y$ and variance $\sigma_x^2 + \sigma_y^2$. We see this in the Brownian motion, since for $s < t$ we can write

$$B_t = B_s + [B_t - B_s].$$

We have written B_t, a normal mean zero, variance t random variable, as the sum of two indepedent mean zero random variables with variances s and $t - s$, respectively. In particular,

$$\mathbf{E}[(B_{t+h} - B_t)^2] = \mathrm{Var}[B_{t+h} - B_t] = h.$$

In other words, in a small time interval of length h the square of the increment, $B_{t+h} - B_t$, is of order h, and the length of the increment is of order $\sqrt{h}$. As $h \to 0$, $\sqrt{h} \to 0$, so this indicates that B_t is a continuous function of t. In fact it can be shown that the Brownian motion can be defined so that $t \to B_t$ is a continuous function.

How smooth is the random function

$$t \to B_t?$$

Let us try to take the derivative:

$$\frac{d}{dt} B_t = \lim_{h \to 0} \frac{B_{t+h} - B_t}{h}.$$

If h is small, then $B_{t+h} - B_t$ is of the order of $\sqrt{h}$. But for h small $\sqrt{h}$ is much larger than h. What we see is that this limit does not exist. In fact, it can be shown that the function $t \to B_t$ is *nowhere differentiable.* The paths of Brownian motion are continuous but very rough.

One of the interesting facets of Brownian motion is its scaling properties. An example of a function that scales nicely is the straight line

$$f(t) = ct.$$

If we take a microscope and look near the origin, we will see that this still looks like a line. More mathematically, if we scale space and time by the same factor a and let

$$f_a(t) = c\left(\frac{1}{a}\right)(at),$$

we get the line back again. What happens when we take the microscope to look at the Brownian motion near 0? One thing is that we will have to use a funny microscope that scales time and space differently— if we scale time by a then we scale space by $\sqrt{a}$. Let

$$Y_t = \frac{1}{\sqrt{a}} B_{at}.$$

Then Y_t also looks like a Brownian motion. To see that this is the correct way to scale, note that

$$\mathrm{Var}(Y_t) = \mathrm{Var}\left(\frac{1}{\sqrt{a}} B_{at}\right) = \frac{1}{a}\mathrm{Var}(B_{at}) = \frac{1}{a}(at) = t.$$

The Brownian motion is statistically invariant under this transformation, although it is not invariant on a "path by path" basis — a particular path of the Brownian motion will of course look different when the scaling is done.

One interesting set that arises from the Brownian motion is the zero set,

$$Z = \{t : B_t = 0\}.$$

The inverse image of any closed set under a continuous function is closed; hence $Z = B^{-1}(\{0\})$ is closed. The one-dimensional random walk is recurrent; it is not too difficult to convince oneself that this will be true for the Brownian motion as well. Hence Z is an unbounded set. In other words,

$$\lim_{T\to\infty} \mathbf{P}\{[1,T] \cap Z \neq \emptyset\} = 1.$$

How about near the origin? Let

$$p(\epsilon) = \mathbf{P}\{[\epsilon^2, \epsilon] \cap Z \neq \emptyset\}.$$

If we scale space by ϵ^2, by the scale invariance we can see that

$$p(\epsilon) = \mathbf{P}\{[1, \epsilon^{-1}] \cap Z \neq \emptyset\}.$$

But this goes to 1 as $\epsilon \to 0$, by recurrence of the Brownian motion. What we see is that there are times arbitarily close to zero with B_t equal to 0, as well as points arbitarily close with B_t positive and B_t negative. Topologically, we say that 0 is not an isolated point of Z. One can show in fact that Z has no isolated points.

An example of a set that is topologically equivalent to Z is the Cantor "middle thirds" set. Let

$$A_0 = [0,1],$$

$$A_1 = [0, \tfrac{1}{3}] \cap [\tfrac{2}{3}, 1],$$

$$A_2 = [0, \tfrac{1}{9}] \cap [\tfrac{2}{9}, \tfrac{1}{3}] \cap [\tfrac{2}{3}, \tfrac{7}{9}] \cap [\tfrac{8}{9}, 1],$$

and in general A_n is obtained from A_{n-1} by removing the middle thirds of all the intervals in A_{n-1}. The Cantor set A is defined by

$$A = \bigcap_{n=0}^{\infty} A_n.$$

Note that the approximation A_n consists of 2^n intervals each of length 3^{-n}, One can check that A also has no isolated points.

A way to distinguish "sizes" of two sets such as the zero set of Brownian motion and the Cantor set is fractal dimension. There are several different way to assign a dimension to a set that allow for fractional dimensions. The two most used are box dimension and Hausdorff dimension, the first being more intuitive and generally easier to compute, while the latter is more useful mathematically. The definitions of fractal dimension differ on some sets, but for the two sets we are considering either notion of dimension will give the same result. We will describe box dimension. Suppose A is a subset of $[0,1]$. Divide the interval into the N subintervals

$$\left[0, \frac{1}{N}\right], \left[\frac{1}{N}, \frac{2}{N}\right], \ldots, \left[\frac{N-1}{N}, 1\right].$$

Let $f(A,N)$ be the number of these intervals which intersect the set A. The box dimension D is the number such that as $N \to \infty$,

$$f(A,N) \approx N^D.$$

More precisely,

$$D = \lim_{N\to\infty} \frac{\ln f(A,N)}{\ln N},$$

assuming the limit exists. In the case of the Cantor set, consider for ease $N = 3^n$. By considering the nth approximation of the Cantor set we can see that

$$f(A, 3^n) = 2^n.$$

Solving the equation

$$(3^n)^D = 2^n$$

gives the value $D = \ln 2/\ln 3 \approx .631$.

It is not so easy to determine the dimension of $Z \cap [0,1]$, the zero set of Brownian motion, and we will content ourselves here with an informal argument. Divide the interval $[0,1]$ into N pieces as above and let X_N be the number of these intervals that intersect Z,

$$X_N = \sum_{j=1}^{N} Y(j,N),$$

where $Y(j,N) = 1$ if

$$Z \cap \left[\frac{j-1}{N}, \frac{j}{N}\right] \neq \emptyset,$$

and $Y(j,N) = 0$ otherwise. Remembering that the expected value of the sum is the sum of the expected values, we see that

$$\mathbf{E}[X_N] = \sum_{j=1}^{N} \mathbf{P}\left\{Z \cap \left[\frac{j-1}{N}, \frac{j}{N}\right] \neq \emptyset\right\}.$$

If we scale time by a factor of $N/(j-1)$, we can see that

$$\mathbf{P}\left\{Z \cap \left[\frac{j-1}{N}, \frac{j}{N}\right] \neq \emptyset\right\} = \mathbf{P}\left\{Z \cap \left[1, 1 + \frac{1}{j-1}\right] \neq \emptyset\right\}.$$

How large is this probability for large j? Recall that in time $1/(j-1)$, the Brownian path tends to go a distance of order $1/\sqrt{j-1}$. Hence in order to cross the origin during this interval we would expect the Brownian motion at time 1 to be at a distance of order $1/\sqrt{j-1}$ from the origin. Since the position is a normal random variable of mean zero and variance one, the probability of being within that distance of the origin should look like a constant times $1/\sqrt{j-1}$. If we believe all this, then we would expect that

$$\mathbf{E}[X_N] \approx (\text{const}) \sum_{j=1}^{N} \frac{1}{\sqrt{j-1}} \approx (\text{const}) N^{1/2}.$$

The last approximation is done by approximating the sum by an integral as we have done before. Since we should expect that $\mathbf{E}[X_N] \approx N^D$, we get

$$D = \frac{1}{2}.$$

This has been far from precise, but it does motivate the correct result that the box (and Hausdorff) dimension of the zero set of Brownian motion is $1/2$. Hence this set is "smaller" than the Cantor set.

Lecture 5

Shuffling and Random Permutations

Consider a deck of N cards which we will label, $1, \ldots, N$. A shuffle or a permutation of the cards is a rearrangement of the cards. More precisely, a permutaion is a one-to-one and onto (i.e., one-to-one correspondence) $\pi : \{1, \ldots, N\} \to \{1, \ldots, N\}$. There are a number of ways of representing permutations. We will represent a permutation π by writing how a deck that starts in its natural order $1, 2, \ldots, N$ looks like after the permutation. For example if $N = 6$, then the permutation

$$\pi_1 = (234156)$$

is the permutation that sends the second card to the first position, the third card to the second position, the fourth card to the third position, etc. You can check that if the deck started in the opposite order, $6, 5, 4, 3, 2, 1$, and the shuffle π_1 were performed then the cards would be in the order $5, 4, 3, 6, 2, 1$. If we start with the natural order $1, 2, 3, 4, 5, 6$ and perform the shuffle π_1 twice we obtain the permutation

$$\pi_1 \circ \pi_1 = (341256).$$

In general we can define a multiplication $\circ$ on permutations by saying that $\pi_1 \circ \pi_2$ is the permutation obtained by first performing π_1 and then performing π_2. For example, if π_1 is as above and

$$\pi_2 = (164532),$$

then

$$\pi_1 \circ \pi_2 = (261543).$$

(You can check this — for example, if the cards start in the natural order, the card number 4 is moved to position 3 by π_1 and the card in position 3 is moved to position 5 by π_2; hence 4 is moved to position 5 after both shuffles are done. So after both permutations, the card numbered 4 has moved to position 5.) It is very important that we write down the order in which we do the shuffles; note that

$$\pi_2 \circ \pi_1 = (645132) \neq \pi_1 \circ \pi_2.$$

The set of permutations on the set $\{1, \ldots, N\}$ is often represented by S_N and is called the symmetric group on N elements. We know that S_N contains $N!$ elements. Its group structure is given by the multiplication $\circ$. The identity permutation ι is the "nonshuffle" that moves no cards. It is easy to check that S_N under $\circ$ satisfies the other conditions for a group: namely, it is associative

$$(\pi_1 \circ \pi_2) \circ \pi_3 = \pi_1 \circ (\pi_2 \circ \pi_3),$$

and for every $\pi \in S_N$ there is an inverse permutation π^{-1} such that

$$\pi \circ \pi^{-1} = \pi^{-1} \circ \pi = \iota \ .$$

We have seen that the multiplication in S_N is not commutative, i.e., the symmetric group is a nonabelian group.

By a random shuffle or a random permutation, we will mean an element chosen from S_N according to some probability distribution. Here are some examples of random shuffles.

Example 1. Random Cut. When one takes a deck of cards and cuts them, one separates the cards into two piles of size k and $N - k$ and then puts the bottom $N - k$ cards on top of the first k. The number k can be any number from 1 to N ($k = N$ corresponds to a "noncut" or the identity permutation). Choose the number k uniformly from $\{1, ..., N\}$ and then do the cut. In other words, we give probabilty $1/N$ to each of the permutations of the form

$$(k+1 \;\; k+2 \;\; \cdots \;\; N \;\; 1 \;\; 2 \;\; \cdots \;\; k).$$

Example 2. Completely Random Shuffle. A completely random shuffle will be a permutation chosen uniformly from S_N; each permutation has probability $1/N!$ of being chosen.

Example 3. Random Transposition. A transposition is a permutation that changes the position of two cards and leaves the remaining cards fixed. It is a nice exercise to show that the symmetric group is generated by transpositions, i.e., if π is any permutation in S_N, then there exist a finite sequence of transpositions $T_1, \ldots, T_j$ such that

$$\pi = T_1 \circ T_2 \circ \cdots \circ T_j.$$

There are $\binom{N}{2} = N(N+1)/2$ transpositions in S_N. A random transposition is obtained by giving each of these transpositions probability $2/N(N+1)$.

Example 4. Random k-Transposition. For $k = 2, 3, \ldots, N$ we define a random k-transposition to be a random permutation chosen by choosing a random number between 1 and k (uniformly between 1 and k) and then transposing the kth card with this card. (if k is chosen then we choose the identity permutation.) Random k-transpositions give a nice computer algorithm for generating completely random shuffles. If one does a random 2-transposition, followed by a random 3- transposition, etc., ending with a random N-transposition, then one can check that one ends up with a permuation chosen uniformly from S_N. This is an efficient algorithm for shuffling cards on a computer. (See Problem 5-4 for a similar algorithm that does *not* give the uniform measure on all permutations.)

Example 5. Riffle Shuffle. The riffle shuffle is a model for what many people do when they actually shuffle cards. A standard riffle shuffle is done by splitting the deck into two pieces and then interlacing the two piles together. Randomness is produced because a shuffler does not always split the deck into exactly equal piles and because the interlacing is not done precisely. (Some magicians can do perfect, nonrandom shuffles, but most of us can do only these random shuffles.) There is a precise model for riffle shuffling. We first split the deck into two piles that are "approximately equal". Let k be the number of cards in the first pile and $N-k$ the number in the second pile. A reasonable model, which turns out to be mathematically very nice, is to choose the size of the piles from the binomial distribution with parameters N and $1/2$; i.e., the probability that k cards are chosen in the first pile is

$$\binom{N}{k} 2^{-N}.$$

Once we have split the deck into piles of size k and $N-k$ we must interlace them. When we interlace the two decks, the order of the first k cards does not change and the order of the last $N-k$ cards do not change. In other words, the permutation is obtained by interlacing

$$1 \;\; 2 \;\; \cdots \;\; k-1 \;\; k$$

with

$$k+1 \;\; k+2 \;\; \cdots \;\; N-k.$$

There are $\binom{N}{k}$ different interlacings that keep the orders. One can see this by noting that this is the number of ways of choosing k positions out of N positions. Once we know which k positions cards $1, \ldots, k$ will occupy, we know the entire permutation.

The result of a riffle shuffle on a deck in the natural order will always be a deck that contains at most two increasing sequences. One possibility is that the cards are not mixed at all. The probability that the riffle shuffle produces the identity permutation is

$$\sum_{k=0}^{N} \mathbf{P}\{k \text{ cards in first pile}\}\mathbf{P}\{\text{identity} \mid k \text{ cards in first pile}\}$$

$$= \sum_{k=0}^{N} \left[\binom{N}{k} 2^{-N}\right] \frac{1}{\binom{N}{k}} = (N+1)2^{-N}.$$

Suppose π is a permutation, other than the identity, which contains only two increasing sequences. Let k be the largest integer such that k is sent to a higher numbered position than $k+1$. Then one can see that the increasing sequences must be $1,\ldots,k$ and $k+1,\ldots,N$. The probability that this was chosen by the riffle shuffle is

$$\mathbf{P}\{k \text{ cards in first pile}\}\mathbf{P}\{\pi \mid k \text{ cards in first pile}\}$$
$$= \left[\binom{N}{k}2^{-N}\right]\frac{1}{\binom{N}{k}} = 2^{-N}.$$

We see that every permutation π, other than the identity, that has a chance of being chosen has the same probability 2^{-N}. This is one of the reasons why this is a nice mathematical model for riffle shuffling.

A random walk on the symmetric group is a sequence of random shuffles each chosen from the same random distribution. More precisely, we let $Y_1, Y_2, \ldots$ be independent random variables taking values in S_N all chosen from the same distribution, and we let Z_n be the permutation

$$\mathbf{Z}_0 = \iota,$$
$$\mathbf{Z}_n = Y_1 \circ Y_2 \circ \cdots \circ Y_n.$$

Let p be be the discrete probability distribution for the Y_i,

$$p(\pi) = \mathbf{P}\{Y_i = \pi\}.$$

Let p_n be the probability distribution for Z_n,

$$p_n = \mathbf{P}\{Z_n = \pi\}.$$

These probability distributions can be computed recursively, at least theoretically, using convolutions. Note that

$$p_n(\pi) = \sum \mathbf{P}\{Z_{n-1} = \pi_{n-1}\}\mathbf{P}\{Y_n = \pi_1\} = \sum p_{n-1}(\pi_{n-1})p_1(\pi_1),$$

where the sums are over all π_1, π_{n-1} such that

$$\pi_{n-1} \circ \pi_1 = \pi_n.$$

Hence

$$\begin{aligned} p_n(\pi) &= \sum_{\pi_{n-1}\circ\pi_1=\pi_n} p_1(\pi_1)p_{n-1}(\pi_{n-1}) \\ &= \sum_{\lambda\in S_N} p_{n-1}(\pi\circ\lambda^{-1})p_1(\lambda) \\ &\doteq (p_{n-1} * p_1)(\pi). \end{aligned}$$

Lecture 6

Seven Shuffles are Enough (Sort of)

In this lecture we will investigate the riffle shuffle or, more precisely, the random walk generated by repeated indepedent riffle shuffles. What we (and card players) would like to know is: how many shuffles are needed to mix the deck up sufficiently? Suppose we start with a deck of 52 cards in the natural order. Let R_n denote the probability distribution on permutations (shuffles) generated by doing n consecutive riffle shuffles. What we would like is for $R_n(\pi)$ to be equal to $1/N!$ for all π, where N is the number of cards in the deck. It turns out that it is impossible to get this exactly, no matter how large one chooses n. However, we would hope that it would be close in some sense. As a measure of closeness we will use *variation distance*: the variation distance between two probabilities P_1 and P_2 on a finite set A is defined by

$$\|P_1 - P_2\| = \frac{1}{2} \sum_{x \in A} |P_1(x) - P_2(x)|.$$

The factor $1/2$ is put in so that the distance is always between 0 and 1. The distance is zero if and only if $P_1 = P_2$, and the distance is always less than or equal to one, since

$$\frac{1}{2} \sum_{x \in A} |P_1(x) - P_2(x)| \leq \frac{1}{2} \left[\sum_{x \in A} P_1(x) + \sum_{x \in A} P_2(x) \right] = 1.$$

A variation distance near one indicates that P_1 and P_2 are significantly different.

Let R_n be the probability distribution obtained from doing the riffle shuffle n times, starting with the cards in the natural order. Let U be the uniform distribution on S_N and let

$$D_n = \|R_n - U\| = \frac{1}{2}\sum_{\pi}\left|R_n(\pi) - \frac{1}{N!}\right|.$$

Note that $D_0 = 1 - (1/N!)$. In theory we could compute $D_1, D_2, \ldots$ by computing $R_1, R_2, \ldots$, using the convolution formula, and then computing D_n directly. However, since there are $N!$ terms in the sum that have to be computed, this is impractical except for very small N.

It will be useful to introduce the notion of an a-shuffle for any positive integer a. Our definition will be such that the riffle shuffle defined in the last lecture will be a 2-shuffle. To do an a-shuffle we divide the deck into a piles of size $k_1, \ldots, k_a$,

$$1 \;\cdots\; k_1,$$

$$(k_1+1) \;\cdots\; (k_1+k_2),$$

$$\vdots$$

$$(k_1+k_2+\cdots+k_{a-1}+1) \;\cdots\; (k_1+k_2+\cdots+k_n) = N.$$

We choose $k_1, k_2, \ldots, k_a$ using the multinomial distribution with parameters N and $a^{-1}, \ldots, a^{-1}$,

$$\mathbf{P}\{k_1 = j_1, \ldots, k_a = j_a\} = \binom{N}{k_1\ k_2\ \cdots k_a} a^{-N} = \frac{N!}{k_1!k_2!\cdots k_a!}a^{-N}.$$

Once the $k_1, k_2, \ldots$ are chosen, the a packs are interlaced in a way that keeps the order of the cards in each of the packs. As in the $a = 2$ case, we will give each of these interlacings the same probability. A simple combinatorial argument shows that the number of such interlacings is given by

$$\binom{N}{k_1\ k_2\ \cdots k_a}$$

and hence, given $k_1, k_2, \ldots, k_n$, the probability that a particular interlacing is chosen is given by

$$\left[\binom{N}{k_1\ k_2\ \cdots k_a}\right]^{-1}.$$

Note that an a-shuffle is well-defined even if a is much bigger than N. In this case almost all of the piles will be empty, but there is nothing forbidding this in the definition.

We now consider a permutation π, and try to find the probability that an a-shuffle results in π. In the previous lecture we did this for $a = 2$. Take a permutation π. Then π consists of r increasing sequences in a natural way. To see this, let $j_1 < j_2 < \ldots < j_s$ be an ordered list of all integers j such that the card currently in position j is sent to a higher position than the card currently in position $j+1$. Then $r = s+1$, and π contains the increasing sequences

$$1\ ,\ 2,\ \ldots\ ,\ j_1,$$

$$j_1 + 1\ , j_1 + 2,\ \ldots\ ,\ j_1 + j_2,$$

$$\vdots$$

$$j_s + 1,\ , j_s + 2,\ \ldots\ ,\ N.$$

In order to get π it is necessary that $a \geq r$. If $a = r$, then the only way to obtain π is to split into piles of size $j_1, \ldots, j_s$ and then interlace to produce π. This has probability

$$\binom{N}{k_1\ k_2\ \cdots k_a} a^{-N} \left[\binom{N}{k_1\ k_2\ \cdots k_a}\right]^{-1} = a^{-N}.$$

If $a > r$, the analysis is a little trickier. We need to split the deck into a piles, such that they can be put together to form s piles as above. By considering the number of places where the deck can be split we can see that the number of allowable splits is given by

$$\binom{N + (a - r)}{N}.$$

By considering each of these possibilities, we can see that the probability that π arises from an a-shuffle is

$$\binom{N+(a-r)}{N} a^{-N}.$$

Note that this formula depends only on a and r, and that the formula is decreasing in r, i.e., the fewer the number of increasing sequences, the larger the probability. You might also check that this formula agrees with our formula found last lecture for $a = 2$, $r = 1$.

A key step of the analysis comes in the following beautiful lemma: *The result of doing an a-shuffle followed by a b-shuffle is an ab-shuffle.* By this we mean that the probability of getting a particular permutation π after doing a b-shuffle followed by an a-shuffle is the same as the probabilty of getting the permuation after an ab-shuffle. In particular, doing n riffle (2-riffle) shuffles is exactly the same as doing one 2^n-riffle shuffle.

This lemma is a little tricky to prove. We will not do it here, but rather refer you to the paper of Mann [**M**] for details. Let us see how the lemma can be used. By the formula for a 2^n-shuffle, the probability that a permutation π arises from n 2-shuffles is

$$\binom{N+2^n-r}{N}(2^n)^{-N},$$

where r denotes the number of increasing sequences in the permutation. Therefore

$$\|R_n - U\| = \frac{1}{2}\sum_{r=1}^{N} A_{N,r}\left|\binom{N+2^n-r}{N}(2^n)^{-N} - \frac{1}{N!}\right|,$$

where $A_{N,r}$ is the number of permuations of N cards with exactly r increasing sequences. It is not easy to calculate the numbers $A_{N,r}$ directly; however, they can be computed recursively (see [**M**] for a formula) in a way that allows easy computer calculation of $\|R_n - U\|$ for small n.

Upon doing this calculation, we see that the variation distance for a regular deck of cards ($N = 52$) is very near 1 for $n = 1, 2, 3, 4$ and then starts dropping rapidly. By $n = 7$, it has take a value between .2 and .3 and by $n = 12$ it is almost zero. The value "seven is enough" is a somewhat arbitrary number for this cutoff phenomenon.

Is variation distance a good measure of distance from randomness? Consider for example a game of bridge. In bridge, four hands of thirteen cards are dealt; generally one player gets the first, fifth, ninth, etc. cards, and similarly for the other players. The order in which a certain player receives his or her cards is irrelevant, so all that is important is the four 13-card hands. Whether or not the deck is close to completely shuffled is not important, only whether the four hands are distributed close to the uniform distribution on all

$$\binom{52}{13\ 13\ 13\ 13}$$

possible ways of dividing 52 cards in four suits. Any probability distribution P on S_N also gives a measure on bridge hands. One can check that the variation distance between the bridge hand measure and the uniform measure is no more than the variation distance between P and the uniform measure on S_N. In this way variation distance is a safe measure of distance. Mann gives an example of a game for which seven shuffles do not suffice; this shows that for some situations a variation distance of more than .2 is too much.

Lecture 7

Markov Chains on Finite Sets

Let S be a finite set which we will call the state space. A (discrete time, time-homogeneous) Markov chain is a sequence of random variables $X_0, X_1, \ldots$ taking values in S such that for some function $P : S \times S \to [0, 1]$ and all $n > 0$ and $x_0, \ldots, x_n \in S$

$$\mathbf{P}\{X_n = x_n \mid X_0 = x_0, \ldots, X_{n-1} = x_{n-1}\}$$
$$= \mathbf{P}\{X_n = x_n \mid X_{n-1} = x_{n-1}\} = P(x_{n-1}, x_n).$$

The first equality expresses the "Markovian" property—the probability of moving to a state x_n depends only on the state of the system at time $n - 1$ and not on the previous states of the system. The second equality expresses the time-homogeneity, the fact that the function P does not depend on n. We let N denote the number of elements of S and we use P to denote the $N \times N$ matrix whose entries are indexed by ordered pairs of elements of S, the (x, y) entry being $P(x, y)$. The matrix P is called the transition matrix for the chain and is a stochastic matrix, i.e., a matrix with nonnegative entries such that the sum of each row equals 1,

$$\sum_{y \in S} P(x, y) = 1, \quad x \in S.$$

Random walks on the symmetric group give an example of a Markov chain. Let $S = S_N$ and let $p(\cdot)$ be a probability distribution on S_N. Then the transition matrix for the chain is given by

$$P(\pi, \lambda) = p(\pi^{-1} \circ \lambda),$$

i.e., the probability that the deck ordered by the permutation π is transformed to a deck ordered by λ is the probability that the shuffle done was $\pi^{-1} \circ \lambda$. For these random walks we gave a convolution formula to determine the probabilities of certain permutations arising after n shuffles. This formula is just a special case of the multiplication rule for transition matrices of Markov chains, which we will now describe. We have already seen that

$$P(x, y) = \mathbf{P}\{X_1 = y \mid X_0 = x\} = \mathbf{P}\{X_n = y \mid X_{n-1} = x\}.$$

Let $P^2(x, y) = \mathbf{P}\{X_2 = y \mid X_0 = x\}$. Then

$$\begin{aligned} P^2(x, y) &= \sum_{z \in S} \mathbf{P}\{X_1 = z, X_2 = y \mid X_0 = x\} \\ &= \sum_{z \in S} \mathbf{P}\{X_1 = z \mid X_0 = x\} \mathbf{P}\{X_2 = y \mid X_0 = x, X_1 = z\} \\ &= \sum_{z \in S} \mathbf{P}\{X_1 = z \mid X_0 = x\} \mathbf{P}\{X_2 = y \mid X_1 = z\} \\ &= \sum_{z \in S} P(x, z) P(z, y). \end{aligned}$$

We see that $P^2(x, y)$ is the (x, y) entry of the matrix $P^2 = P \cdot P$ (which makes the notation very convenient!). We can continue this argument and see that

$$\mathbf{P}\{X_n = y \mid X_0 = x\} = P^n(x, y),$$

where $P^n(x, y)$ denotes the (x, y) entry of the matrix P^n. This formula is even true when $n = 0$, since $P^0 = I$, the identity matrix. If the state space S is small, and the transition matrix P is given, we can determine P^n easily using a computer. For large state spaces, such as the symmetric group S_{52}, it is impossible to do the matrix calculations directly on a computer.

What happens when we run a Markov chain for a long time? Algebraically, this is the same thing as asking what does the matrix

P^n look like for large n. What we would hope is that the system settles down to a particular probability distribution, i.e., that there is a probability measure m on S such that

$$\lim_{n\to\infty} P^n(x,y) = m(y),$$

or in matrix notation

$$\lim_{n\to\infty} P^n = \begin{bmatrix} \mathbf{m} \\ \mathbf{m} \\ \vdots \\ \mathbf{m} \end{bmatrix}.$$

Here $\mathbf{m}$ denotes the N-vector with components $m(x)$. (We will use boldface to denote vectors; whether the vector is to be treated as a row vector or a column vector must be determined by context.) Any such $\mathbf{m}$ satisfies

$$\begin{bmatrix} \mathbf{m} \\ \mathbf{m} \\ \vdots \\ \mathbf{m} \end{bmatrix} = \lim_{n\to\infty} P^n = \big[\lim_{n\to\infty} P^{n-1}\big] P = \begin{bmatrix} \mathbf{m} \\ \mathbf{m} \\ \vdots \\ \mathbf{m} \end{bmatrix} P,$$

and hence

$$\mathbf{m}P = \mathbf{m}. \tag{7.1}$$

We call a probability vector $\mathbf{m}$ (a probability vector is a vector with nonnegative components whose sum is 1) an invariant probability if it satisfies (7.1). Other terms often used are steady-state probability, stationary probability, or equilibrium probability. The invariant probability is a left eigenvector for the matrix P with eigenvalue 1. For a simple example, consider the chain with $S = \{1, 2\}$ and

$$P = \begin{bmatrix} 1-p & p \\ q & 1-q \end{bmatrix},$$

where $0 < p, q < 1$, The equation $\mathbf{m}P = \mathbf{m}$ leads to the equations

$$\begin{aligned} (1-p)m(0) + qm(1) &= m(0), \\ pm(0) + (1-q)m(1) &= m(1). \end{aligned}$$

These equations are redundant and any vector of the form (qa, pa), a real, is an eigenvector. Imposing the condition $m(0) + m(1) = 1$ chooses a unique eigenvector

$$\mathbf{m} = \left(\frac{q}{p+q}, \frac{p}{p+q} \right).$$

One can check for particular values of p, q that approaches the matrix with rows $\mathbf{m}$. In particular, if $p = q$,

$$\lim_{n \to \infty} P^n = \begin{bmatrix} 1/2 & 1/2 \\ 1/2 & 1/2 \end{bmatrix}.$$

For $p = q$, the limit does not depend on the particular value of p; however, the rate of convergence does. If p is near 0 or 1, the convergence is slow while for p in the middle of the unit interval it is quite fast.

If $\mathbf{1}$ denotes the vector $(1, 1, \ldots, 1)$, then any stochastic matrix P satistifies

$$P\mathbf{1} = \mathbf{1},$$

i.e., $\mathbf{1}$ is a right eigenvector with eigenvalue 1. This implies that there is a left eigenvector with eigenvalue 1. If such a nonzero eigenvector can be chosen with all nonnegative entries, then it can be normalized so that it is a probability vector. It is not so easy to show that the eigenvector can be chosen with all nonnegative entries. There is a theorem in linear algebra, called the Perron-Froebenius Theorem, that states that if the stochastic matrix P has all *positive* entries, then the eigenvector is unique (up to a scalar); can be chosen will all positive entries; and hence the scalar can be chosen so that the vector is a probability vector. Moreover, the Perron-Froebenius Theorem tells us that all other eigenvalues have (complex) absolute values strictly less than 1. So stochastic matrices with all positive entries have a unique invariant probability $\mathbf{m}$. Many interesting examples, such as the random walk arising from the riffle shuffle, do not satisfy the property that P has all positive entries. However, every permutation can be realized by a sequence of six riffle shuffles (why?). Hence for this P, P^6 has all positive entries. The matrix P^6 satisfies the conditions of the Perron-Froebenius Theorem and hence there exists a unique $\mathbf{m}$

with

$$\mathbf{m}P^6 = \mathbf{m},$$

and all other eigenvalues of P^6 are strictly less than one in absolute value. From this, with a little thought, we can deduce that $\mathbf{m}P = \mathbf{m}$ and all other eigenvalues of P have absolute value strictly less than one. By a similar argument, we can show that if P is any stochastic matrix with the property that for some n, P^n has all positive entries, then there exists a unique invariant probability vector $\mathbf{m}$, and all other eigenvalues have absolute values strictly less than one.

Not every stochastic matrix has the property that P^n has all positive entries for some n. Suppose that the state space S is partitioned into two sets S_1 and S_2 such that states in S_1 do not change into states in S_2 and vice versa, i.e., such that $P(x,y) = 0$ for $x \in S_1, y \in S_2$ or $x \in S_2, y \in S_1$. Then it is easy to see that $P^n(x,y) = 0$ for any n and any $x \in S_1, y \in S_2$. Such chains are called reducible; chains such that you can get from any one state to any other state in some number of steps are called irreducible. Another obstruction to having all positive entries is periodicity—we saw this phenomenon in the simple random walk. Suppose, for example, that the state space is divided into two sets S_1, S_2 (say "even" and "odd" points) such that one always moves from S_1 to S_2 or from S_2 to S_1. So, $P(x,y) = 0$ if $x, y \in S_1$ or $x, y \in S_2$. Then one can check that $P^{2n}(x,y) = 0$ for $x \in S_1, y \in S_2$. Chains that do not have such a problem with periodicity are called aperiodic. It is a nice exercise to check that any irreducible chain that has a point x with $P(x,x) > 0$ is aperiodic. For any irreducible, aperiodic chain there is an n such that P^n has all positive entries.

We will now sketch the proof of one of the fundamental theorems of Markov chains: if P is the transition matrix for an irreducible, aperiodic chain, then

$$\lim_{n\to\infty} P^n = \begin{bmatrix} \mathbf{m} \\ \mathbf{m} \\ \vdots \\ \mathbf{m} \end{bmatrix}.$$

We will do the proof in the case when the matrix P can be diagonalized (the case when the matrix cannot be diagonalized can be handled similarly using the Jordan canonical form). We write

$$D = Q^{-1}PQ, \quad P = QDQ^{-1}.$$

The matrix D is a diagonal matrix with entries $\lambda_1, \ldots, \lambda_N$, the eigenvalues of P. We can choose Q so that the eigenvalues appear on the diagonal in decreasing order,

$$1 = \lambda_1 > |\lambda_2| \geq |\lambda_3| \geq \cdots \geq |\lambda_N|.$$

The columns of Q are the right eigenvectors and the rows of Q^{-1} are the left eigenvectors. By choosing $\mathbf{1}$ as the right eigenvector and $\mathbf{m}$ as the left eigenvector (for the eigenvalue 1) we have

$$P = QDQ^{-1}$$
$$= \begin{bmatrix} 1 & * & \cdots & * \\ 1 & * & \cdots & * \\ & \vdots & & \\ 1 & * & \cdots & * \end{bmatrix} \begin{bmatrix} 1 & 0 & \cdots & 0 \\ 0 & \lambda_2 & \cdots & 0 \\ & \vdots & & \\ 0 & 0 & \cdots & \lambda_N \end{bmatrix} \begin{bmatrix} m(1) & m(2) & \cdots & m(N) \\ * & * & \cdots & * \\ & \vdots & & \\ * & * & \cdots & * \end{bmatrix}.$$

Since $P^n = (QDQ^{-1})^n = QD^nQ^{-1}$, it follows that P^n is equal to

$$\begin{bmatrix} 1 & * & \cdots & * \\ 1 & * & \cdots & * \\ & \vdots & & \\ 1 & * & \cdots & * \end{bmatrix} \begin{bmatrix} 1 & 0 & \cdots & 0 \\ 0 & \lambda_2^n & \cdots & 0 \\ & \vdots & & \\ 0 & 0 & \cdots & \lambda_N^n \end{bmatrix} \begin{bmatrix} m(1) & m(2) & \cdots & m(N) \\ * & * & \cdots & * \\ & \vdots & & \\ * & * & \cdots & * \end{bmatrix}.$$

Since the eigenvalues $\lambda_2, \ldots, \lambda_N$ have absolute value less than one, $\lambda_i^n \to 0,\ i > 1$. Therefore,

$$\lim_{n\to\infty} P^n = \begin{bmatrix} 1 & * & \cdots & * \\ 1 & * & \cdots & * \\ & \vdots & & \\ 1 & * & \cdots & * \end{bmatrix} \begin{bmatrix} 1 & 0 & \cdots & 0 \\ 0 & 0 & \cdots & 0 \\ & \vdots & & \\ 0 & 0 & \cdots & 0 \end{bmatrix} \begin{bmatrix} m(1) & m(2) & \cdots & m(N) \\ * & * & \cdots & * \\ & \vdots & & \\ * & * & \cdots & * \end{bmatrix}$$
$$= \begin{bmatrix} m(1) & m(2) & \cdots & m(N) \\ m(1) & m(2) & \cdots & m(N) \\ & \vdots & & \\ m(1) & m(2) & \cdots & m(N) \end{bmatrix}.$$

We also see from this analysis that the rate of convergence is determined by the size of the other eigenvalues; in particular, the size of the second largest eigenvalue λ_2 is critical in determining the rate of convergence.

As an example, consider the 2-state Markov chain with $p = q < 1/2$. Then $\mathbf{m} = (1/2, 1/2)$, and the second eigevalue is $1 - 2p$ with eigevector $(1, -1)$. Note that if p is near zero, the second eigenvalue is very near one and it takes a long time for the chain to reach its equilbrium state. However, if p is near $1/2$, the chain reaches equilibrium quickly. It is often possible to estimate the size of the second eigenvalue for chains with a large number of states. In the case of the riffle shuffle, the second largest eigenvalue is $1/2$ (see [**M**]). The fact that this is not very close to 1 tells us that the chain reaches equilibrium quickly.

Finding the invariant probability for an N-state Markov chain can be difficult if N is large. But there are some cases where it is easy. Suppose P is irreducible, aperiodic, and *symmetric*,

$$P(x, y) = P(y, x), \quad x, y \in S.$$

Then the invariant probability is the uniform measure which gives probability $1/N$ to each state. To see this it suffices to show that the uniform probability vector is a left eigenvector (since the invariant probability is unique). But left eigenvectors and right eigenvectors are the same for symmetric matrices, and we have already seen that vectors with all entries the same are right eigenvectors of stochastic matrices. More generally, suppose there is a probability measure $\mathbf{m}$ on S such that for all x, y,

$$P(x, y)m(x) = P(y, x)m(y). \tag{7.2}$$

We say P is *reversible* with respect to $\mathbf{m}$ if this holds. Note that if P is reversible, then for every x,

$$\sum_{y \in S} m(y)P(y, x) = \sum_{y \in S} m(x)P(x, y) = m(x) \sum_{y \in S} P(x, y) = m(x),$$

i.e., $\mathbf{m}$ is an invariant probability. If the chain is irreducible and aperiodic, then we can conclude that $\mathbf{m}$ is the unique invariant probability. Reversible Markov chains have the property that in equilibrium they look the same whether time is moving forward or backwards. The random walk associated with the riffle shuffle is an example of a nonreversible Markov chain. In many applications, one can find a positive function m that statisfies (7.2) but that is not necessarily a probability measure. If we multiply m by the appropriate scalar so that it becomes a probability measure, we can see that this must be the invariant probability.

Lecture 8

Markov Chain Monte Carlo

A recent application of Markov chains has been in simulation of random systems on large finite sets. The general framework is as follows: a large finite set S is given with a positive function f defined on the set. The goal is to draw an element at random from the set S, where the set is given the probability measure

$$m(x) = \frac{f(x)}{\sum_{y \in S} f(y)}.$$

Often the denominator is not known. If the set S is very large, it is not clear how to write an algorithm which picks an element at random. Even when f is identically 1, so that m is the uniform probability distribution of S, there may be no efficient algorithm (or even any way to accurately count the number of elements in S).

The idea of Markov Chain Monte Carlo (MCMC) is to find an irreducible, aperiodic Markov chain defined on the set S whose invariant probability is m. If the chain is relatively easy to simulate on the computer, then the theorem described in the previous lecture gives an algorithm: start with some $x \in S$; run the chain sufficiently long so that the chain is nearly in equilibrium; then choose the value of the chain as your value. We are used to doing this when we shuffle cards: when we want to choose a random element from the uniform distribution on S_N, we perform a Markov chain (e.g., random walk using riffle shuffles), until we think that the deck is approximately uniform.

As an example, consider the set $T = T_K$ of all $K \times K$ matrices of 0s and 1s such that no two 1s are next to each other, either in a row or in a column. More formally, T is the set of $K \times K$ matrices $M = [M(i,j)]$ such that

$$M(i,j) \in \{0,1\}, \quad 1 \leq i,j \leq K,$$

and for every (i,j), if $M(i,j) = 1$ then

$$M(i-1,j) = M(i+1,j) = M(i,j-1) = M(i,j+1) = 0.$$

(Sometimes "periodic boundary conditions" are given, but we will not worry about this here.) Consider the uniform measure on all such matrices. There is no closed formula for the number of elements of T, and it is not clear how to choose a matrix so that all matrices have the same chance of being chosen. Consider, however, the following algorithm: take some $M \in T$; choose an entry of M at random, uniformly among the K^2 entries of T; if the entry is a 1, change the entry to a 0; if the entry is a 0, change it to a 1 unless this would result in a matrix that is not in T—if the change is not permitted then make no change to the matrix M; and output the matrix we now have (which is either the matrix we started with or a matrix with one entry changed). This procedure can be considered as a Markov chain with state space T, and transition matrix

$$P(M_1, M_2) = \frac{1}{K^2},$$

if $M_1, M_2 \in T$ are two legal matrices that differ in exactly one entry; $P(M_1, M_2) = 0$, if M_1 and M_2 differ in more than one entry; and

$$P(M_1, M_1) = \frac{j}{K^2},$$

where j is the number of entries of M_1 that are 0 and such that changing the entry to a 1 would result in a matrix that is not in T. It is not difficult to check that this gives an irreducible, aperiodic Markov chain that is symmetric,

$$P(M_1, M_2) = P(M_2, M_1).$$

Hence, the uniform distribution is the invariant probability for this chain. If we start with any matrix $M \in T$ and run the chain for a sufficiently long time, then the matrix obtained at that time will have an almost uniform distribution.

This example can be considered as a special case of a large class of models in equilibrium statistical physics. Let $S = S(K)$ be the set of all $K \times K$ matrices of 0s and 1s. (Often physicists take matrices of -1s and $+1$s instead, considering a $+1$ as a spin upwards and a -1 as a spin downwards.) Let g be a symmetric function $g : \{0, 1\} \times \{0, 1\} \to \mathbf{R}$; such a function is determined by three values: $g(0, 0)$, $g(0, 1) = g(1, 0)$, and $g(1, 1)$. Let $\beta > 0$ be a parameter measuring the strength of some interaction. We define the energy of the matrix M to be

$$E = \sum_{(i,j)\sim(i',j')} g(M(i, j), M(i', j')),$$

where $(i, j) \sim (i', j')$ indicates that the sum is over all "nearest neighbors", i.e., all pairs (i, j),(i', j') with

$$|i - i'| + |j - j'| = 1.$$

In equilibrium statistical physics it is natural to look at a particular probability measure on T that gives equal weight to all matrices with the same energy. Let

$$f(M) = f_\beta(M) = e^{-\beta E}.$$

Note that f is large when the energy E is small. The previous example is a limiting case where $g(0, 0) = g(0, 1) = 0$, $g(1, 1) = 1$ and $\beta = \infty$. One important example is the Ising model, for which $g(0, 0) = g(1, 1) = 0$ and $g(0, 1) = 1$; in the Ising model, configurations (matrices) with many pairs of neighboring spins pointing in opposite directions have high energy.

Suppose M_1 and M_2 are matrices that differ in only one entry. Then one can calculate

$$\frac{f(M_2)}{f(M_1)},$$

with a simple calculation looking only at the nearest neighbors of the entry being changed. Consider the following algorithm: start with a matrix $M_1 \in S$; choose an entry at random choosing uniformly among all K^2 entries and let M_2 be the matrix obtained by changing the chosen entry; if $f(M_2)/f(M_1) \geq 1$, return the matrix M_2. If $f(M_2)/f(M_1) = q < 1$, then with probability q return the matrix M_2 and with probability $1 - q$ return the matrix M_1. This corresponds to the Markov chain whose transition matrix P is given as follows: if M_1, M_2 differ in more than one entry,

$$P(M_1, M_2) = 0;$$

if M_1, M_2 differ in exactly one entry,

$$P(M_1, M_2) = \frac{1}{K^2} \min\left\{1, \frac{f(M_2)}{f(M_1)}\right\};$$

and $P(M_1, M_1)$ is whatever is required so that

$$\sum_{M \in S} P(M_1, M) = 1.$$

Again it is easy to check that this chain is irreducible and (except in trivial cases) aperiodic. Also note that

$$f(M_1)P(M_1, M_2) = f(M_2)P(M_2, M_1).$$

Hence this chain is reversible with respect to f and has invariant measure

$$m(M) = Z^{-1} f(M),$$

where

$$Z = \sum_{M_1 \in S} f(M_1).$$

As another example consider Ω_N, the set of self-avoiding walks of length N in $\mathbf{Z}^2$ as described in Lecture 3. As we have already noted, there is no nice expression for C_N, the cardinality of Ω_N. Still the uniform probability measure on Ω_N is a natural model for polymer chains, so we would like to be able to sample from this distribution. There seems to be no efficient method for using a computer to generate an SAW of length N under the constraint that all SAWs have exactly the same chance of being selected. There are some MCMC methods which do allow selection of SAWs from distributions very close to the uniform distribution, and some of the best simulations of

SAWs have been made using these methods. The one we will describe here is called the *pivot algorithm*. Let $\mathcal{O}$ denote the set of orthogonal transfomations in the plane that fix the origin and map the square lattice onto itself. The set $\mathcal{O}$ consists of rotations of $\pi/2, \pi, 3\pi/2$ and reflections about coordinate axes and the diagonals $y = x, y = -x$. Here is the algorithm: start with an $\omega = (\omega_0, \cdots, \omega_N) \in \Omega_N$; choose a number at random from $0, \ldots, N-1$, using the uniform distribution, and call this number k; choose a transformation T from $\mathcal{O}$ according to the uniform distribution on this finite set; consider the walk obtained by fixing the first k steps of the walk, but performing the transformation T on the remaining part of the walk, using ω_k as the origin for the transformation; after doing this transformation, we get a new simple random walk path that may or may not be self-avoiding; if it is self-avoiding, choose that as the new walk, and if it is not just return the walk that we started with. We can see that this gives a well-defined Markov chain on Ω_N whose transition matrix P is symmetric,

$$P(\omega_1, \omega_2) = P(\omega_2, \omega_1).$$

It is not immediately obvious that the chain is irreducible, but one can check that any SAW can be unwrapped to form a completely straight SAW, and hence one can go from any SAW to any other SAW through a series of transformations. Also, $P(\omega, \omega) > 0$ for all ω, and hence the chain is aperiodic.

The theorem now tells us that the Markov chain converges to the equilibrium measure which is the uniform measure on paths. One might expect that the rate of convergence would be very slow because most transformations are rejected. In fact, it is conjectured that the fraction of accepted transformations looks only like N^{-a} for some $a > 0$ which is not too big. Once a transformation is made, the SAW moves a lot, so not too many successful transformations are needed to get to an approximately uniform distribution. It is expected (but has not been proven) that this chain gets uniform in polynomial time, i.e., like some power of N, and the exponent is not very large. Actual simulations using this algorithm are consistent with this conjecture. An algorithm that takes time N^b to produce one SAW may not seem very efficient, but it is much better than naive algorithms which will take an exponential (in N) amount of time to produce one SAW.

Let us return to our first example, matrices of 0s and 1s with neighboring 1s forbidden. Let us consider two possible questions that we could ask:

— What is the density (fraction) of 1s in a typical such matrix?

— How many such matrices are there?

The first question is very well suited for numerical simulation; run the Markov chain for a long time and keep track of the fraction of 1s in each matrix (in actually doing this simulation, one will find that the convergence is slightly quicker for the problem if one imposes periodic boundary conditions, i.e., makes entry $(1,1)$ and $(1,K)$ neighboring entries, etc.). The second question is not so easy to handle. The algorithm is designed to sample from the uniform distribution without knowing how many elements there are; in fact, this is one of the beauties of the algorithm. However, there is a way to get an approximation of the number of legal matrices. Fix K, and let $T = T_K$ be as defined before. Let $A = \{(i,j) : 1 \leq i,j \leq K\}$, and suppose that

$$\emptyset = A_0 \subset A_1 \subset A_2 \subset \cdots \subset A_m = A.$$

Consider the set W_k of all matrices M in T_K such that $M(i,j) = 0$ for all (i,j) in $A \setminus A_k$. Let $\#(W_k)$ denote the number of elements in W_k. Then

$$\#(T) = \#(W_m) = \frac{\#(W_1)}{\#(W_0)} \frac{\#(W_2)}{\#(W_1)} \cdots \frac{\#(W_m)}{\#(W_{m-1})}.$$

For each k, it is easy to adapt the Markov chain so that only matrices in W_k are chosen; we just forbid all moves that take us to a matrix outside of W_k. By running the Markov chain, for a long time we can then estimate the fraction of legal matrices in W_k that are actually in W_{k-1}. This gives an estimate of

$$\frac{\#(W_{k-1})}{\#(W_k)}.$$

If we do this for each k we can get a rough approximation of $\#(T)$. See Lecture 12 for some results of such a simulation.

Lecture 9

Random Walks and Electrical Networks

In this lecture we discuss the relationship between random walks on weighted graphs and electrical networks. This is an example of two different ideas producing essentially the same mathematical model. Since the model is the same, we get the advantage of using both random walk and electrical network intuition in trying to analyze the mathematical model, regardless of whether we were originally interested in random walks or electrical networks.

We start with random walks on weighted graphs. We will let $G = (S, E)$ denote a finite simple (undirected) graph, i.e., a finite set of vertices S and a collection of edges $E = \{e_1, \dots, e_m\}$. Each edge connects two different vertices, and we say two vertices are *adjacent* if there is an edge connecting them. Equivalently, we can think of edges as unordered pairs of distinct vertices. The term *simple* implies that there is at most one edge connecting any pair of vertices. We call the graph *connected* if for any two vertices one can find a path of edges connecting them, i.e., if for every $x, y \in S$ there exist vertices $z_0, \dots, z_k$ with $z_0 = x, z_k = y$ and z_i adjacent to z_{i-1} for $1 \leq i \leq k$. A *weighted* (simple) graph is a simple graph $G = (S, E)$ along with a function

$$w : E \to (0, \infty).$$

Since edges are unordered pairs of vertices, we can also think of w as a function from $S \times S$ to $[0, \infty)$ that is symmetric ($w(x, y) = w(y, x)$) and such that $w(x, x) = 0$ for all x. The set of edges can then be defined by

$$E = \{\{x, y\} : w(x, y) > 0\}.$$

Weighted graphs are sometimes called (undirected) networks. We let $d(x)$ be the total weight of edges out of the vertex x,

$$d(x) = \sum_{y \in S} w(x, y).$$

If $G = (S, E, w)$ is a weighted graph, there is a Markov chain, called *random walk* on the weighted graph with state space S, defined by the transition matrix

$$P(x, y) = \frac{w(x, y)}{d(x)},$$

assuming $d(x) > 0$. If $d(x) = 0$, we set $P(x, x) = 1$. For ease, we will assume unless stated otherwise that $d(x) > 0$ for each x. It is easy to check that the chain is irreducible if and only if the graph G is connected. If the chain is irreducible, it can have period 1 or period 2. It is easy to see that the period can be no more than two, since $P^2(x, x) > 0$ for every x (why?). An irreducible chain has period 2 if and only if the graph is *bipartite*, i.e., the set of vertices can be divided into two disjoint sets S_1, S_2 such that every edge connects a vertex in S_1 to a vertex in S_2. Note that

$$d(x)P(x, y) = d(y)P(y, x).$$

Hence the chain is reversible and has invariant probability

$$m(x) = \frac{d(x)}{\sum_{y \in S} d(y)}.$$

An electrical network is very similar to a weighted graph as described above. In an electrical network we have a finite, simple graph $G = (S, E)$ (this discussion can easily be extended to infinite and non-simple graphs, but we will conisder only finite, simple graphs here) and a function

$$R : E \to (0, \infty)$$

that is called the *resistance* between the vertices $\{x, y\}$. We define the *conductance* to be the reciprocal of the resistance. The conductance is the quantity that corresponds to the weight above, so we will let

$$w(x) = \frac{1}{R(x)}.$$

Any electrical network corresponds to a weighted graph, and vice versa; we just have to remember that the weight of an edge is the reciprocal of the resistance of the edge. Pairs of vertices that are not connected can be considered as edges that have zero weight, or equivalently as edges that have infinite resistance.

It is standard in electrical network theory to fix the voltages on a subset of the vertices and to consider the effect on the remaining voltages. Suppose a finite set of vertices $B \subset S$ is chosen and a voltage V is specified on the vertices in B. (One example is the case when B is just two points, say x, y. In this case it is often the voltage drop $V(x) - V(y)$ that is specified, but for ease we will assume that the exact voltage is given on B.) The goal is to find the voltage at each vertex in S. Let $C = S \setminus B$ be the set of points for which the voltage is not specified; these are often called the *interior* vertices while B contains the *boundary* vertices. Kirchhoff's laws imply that the *current* going into any interior vertex is the same as the current going out of the vertex, where the current $I(x, y)$ going from vertex y to vertex x is

$$I(x, y) = \frac{V(y) - V(x)}{R(x, y)} = w(x, y)[V(y) - V(x)].$$

Note that $I(x, y) = -I(y, x)$. Kirchhoff's law then implies that for $x \in C$,

$$\sum_{y \sim x} I(x, y) = 0,$$

i.e.,

$$\sum_{y \sim x} w(x, y)[V(y) - V(x)] = 0. \tag{9.1}$$

Here $x \sim y$ means that $\{x, y\} \in E$.

A function V satisfying (9.1) is said to be *w-harmonic* at x. We let $\Delta = \Delta_w$ be the operator on functions on the vertices (an operator on functions here is just a function on the set of functions, i.e., a rule that specifies for each function on the vertices another function on the vertices):

$$\Delta f(x) = \sum_{y \sim x} w(x,y)[V(y) - V(x)] = \left[\sum_{y \sim x} w(x,y)V(y)\right] - d(x)V(x).$$

A function f is called w-harmonic at x if $\Delta f(x) = 0$. The problem from electrical network theory can be stated as follows: given a function $V : B \to \mathbf{R}$, find a function $V : S \to \mathbf{R}$ that agrees with V on B and that is w-harmonic on C, i.e., for each $x \in C$,

$$\Delta V(x) = 0. \tag{9.2}$$

This is an example of what mathematicians call a *Dirichlet problem.* Note that (9.2) gives k linear equations in k unknowns, where k is the number of elements in C. The unknowns are $V(x), x \in C$, since we are assuming that the values $V(x), x \in B$, are given. From what we know about linear equations we would guess that there is a unique solution to the system of equations, but of course there might be a chance that this is not true. As it turns out, there *is* a unique solution to the problem, and the random walk gives us a nice way of describing the solution.

We will assume that the graph is connected, and let X_n denote the position of the random walk on the network. We let $T = T(B)$ be the (random) time

$$T = \min\{j \geq 0 : X_j \in B\}.$$

It is easy to check that, since the graph is connected (and hence the chain is irreducible), with probability one $T < \infty$ no matter which vertex we start at. If we start at a vertex in B, then $T = 0$, while if we start in C, $T > 0$. Assume V is defined on B. We then define V on all of S by the formula

$$V(x) = \mathbf{E}[V(X_T) \mid X_0 = x].$$

Since $T = 0$ if $x \in B$, this definition agrees with V on B. Suppose $x \in C$. Then the Markov property implies that

$$\begin{aligned} V(x) &= \sum_{y \in S} P(x,y)\mathbf{E}[V(X_T) \mid X_0 = x, X_1 = y] \\ &= \sum_{y \in S} \frac{w(x,y)}{d(x)} V(y), \end{aligned}$$

and hence

$$\left[\sum_{y \in S} w(x,y)V(y)\right] - d(x)V(x) = 0.$$

This shows that V is w-harmonic. This gives *existence* of a solution of the Dirichlet problem; we would also like to show *uniqueness*. To do this, suppose V were defined on B and we had two functions V_1, V_2 defined on S with

$$V_1(x) = V_2(x) = V(x), \quad x \in B,$$

$$\Delta V_1(x) = \Delta V_2(x) = 0, \quad x \in C.$$

In particular, if we let $f(x) = V_1(x) - V_2(x)$ we see that

$$f(x) = 0, \quad x \in B,$$

$$\Delta f(x) = 0, \quad x \in C.$$

Choose a point x in C such that $f(x)$ is maximum. Then one can check that the condition $\Delta f(x) = 0$ implies that every vertex y adjacent to x must satisfy $f(y) = f(x)$, since a single y with $f(y) < f(x)$ would imply that $\Delta f(x) < 0$. We can continue this argument for all the points adjacent to points adjacent to x, etc., and we would conclude that for all z, $f(z) = f(x)$. This gives $V_1(x) = V_2(x)$ for all x, and hence the solution is unique.

We have given a probabilistic solution of a system of linear equations. Sometimes the system of equations can tell us something about the random walk. Suppose $B = B_1 \cup B_2$ for disjoint B_1, B_2, and let $T = T(B)$ be as in the previous paragraph. Suppose we are interested in the probability that the random walk on the graph reaches the set B_1 before it reaches B_2. We can write this probability as the function

$$f(x) = \mathbf{P}\{X_T \in B_1 \mid X_0 = x\} = \mathbf{E}[V(X_T)],$$

where V is the function that equals 1 on B_1 and 0 on B_2. Then this function is the unique function f which agrees with V on B and is such that, for each $x \in C$,

$$d(x)f(x) = \sum_{y \in S} w(x,y)f(y).$$

As an example let $S = \{0, 1, \ldots, N\}$ and let $w(j, j+1) = 1/2, j = 0, 1, \ldots, N-1$. Random walk on this directed graph is the same as usual simple random walk, at least until the walker reaches one of the boundary points $B = \{0, N\}$. Let $B_1 = \{N\}, B_2 = \{0\}$. The the probability $f(x)$ that a simple random walker starting at a vertex x reaches N before reaching 0 is given by the unique function f satisfying $f(0) = 0, f(N) = 1$, and

$$f(j) = \frac{1}{2}f(j-1) + \frac{1}{2}f(j+1), \quad 1 \leq j \leq N-1.$$

There are a number of ways to determine the solution. We will just pull it out of the hat and say that the solution is

$$f(j) = \frac{j}{N}.$$

It is easy to check that this satisfies the equation above, and the uniqueness of the solution then implies that this must be the appropriate probability.

We will now consider an example where the electrical network perspective helps us discover something about the random walk. To do this we introduce the notion of *effective resistance* between a vertex a and a set of vertices B_1. The effective resistance is the amount of voltage needed to be placed on the vertices of B_1 so that if zero voltage is put on a, a current of one is produced. Let V be the voltages: $V(x) = V_0$ at B_1 (V_0 to be determined) and $V(a) = 0$. Let $B = \{a\} \cup B_1$ and, as before, $C = S \setminus B$. At each $x \in C$ the voltage satisfies

$$\sum_{y \sim x} w(x,y)[V(y) - V(x)] = 0.$$

The fact that a unit current is flowing can be expressed by saying that the total current into a is 1:

$$\sum_{y \sim a} w(a,y)V(y) = 1.$$

One might also express this fact by saying that the total current flowing out of B_1 is 1:

$$\sum_{b \in B_1} \sum_{y \sim b} w(y,b)[V(b) - V(y)] = 1. \tag{9.3}$$

It is not immediately obvious that these are equivalent statements. However, we can check this as follows. First note that

$$\sum_{x \in S} \sum_{y \in S} w(x,y)[V(y) - V(x)] = 0,$$

since for any pair of vertices $\{x, y\}$,

$$w(x,y)[V(y) - V(x)] = -w(y,x)[V(x) - V(y)].$$

Also, since V is w-harmonic at each $x \in C$,

$$\sum_{x \in C} \left(\sum_{y \in S} w(x,y)[V(x) - V(y)] \right) = 0.$$

Therefore,

$$\begin{aligned} 0 &= \sum_{x \in B} \sum_{y \in S} w(x,y)[V(y) - V(x)] \\ &= \left(\sum_{y \in S} w(a,y)V(y) \right) - \sum_{x \in B_1} \sum_{y \in S} w(x,y)[V(x) - V(y)]. \end{aligned}$$

We will use summation by parts to get another formula for the effective resistance. Let

$$\mathcal{E} = \mathcal{E}(V, w) = \frac{1}{2} \sum_{x \in S} \sum_{y \in S} w(x,y)[V(y) - V(x)]^2.$$

Note that

$$\begin{aligned}\mathcal{E} &= \frac{1}{2}\sum_{x\in S}\sum_{y\in S} w(x,y)[V(y)-V(x)]V(y) \\ &\quad -\frac{1}{2}\sum_{x\in S}\sum_{y\in S} w(x,y)[V(y)-V(x)]V(x) \\ &= \sum_{y\in S} V(y)\sum_{x\in S} w(x,y)[V(y)-V(x)].\end{aligned}$$

Split the last sum into the sum over y in C and the sum over y in B. Since V is w-harmonic on C,

$$\sum_{x\in C} w(x,y)[V(y)-V(x)] = 0.$$

Also, $V(a) = 0$ and $V = V_0$ on B_1. Hence,

$$\mathcal{E} = \sum_{y\in B_1} V_0 \sum_{x\in S} w(x,y)[V(y)-V(x)] = V_0.$$

The last equation follows from (9.3). If we let $\tilde{V} = V_0^{-1}V$ be the voltage obtained by setting voltage 0 at a and voltage 1 on B_1, then we see that

$$\mathcal{E}(\tilde{V}, w) = \frac{1}{V_0}.$$

In other words the effective resistance is given by

$$\text{effective resistance} = \frac{1}{\mathcal{E}(\tilde{V}, w)}$$

where $\tilde{V}$ is the w-harmonic function with boundary values 0 at a and 1 on B_1.

We now claim that

$$\mathcal{E}(\tilde{V}, w) = \inf \mathcal{E}(f, w),$$

where the infimum is over all (not necessarily w-harmonic) functions f satisfying the boundary conditions, $f(a) = 0; f(x) = 1, x \in B_1$. This turns out to be a simple exercise in multivariable calculus. By differentiating, we see that the function that attains the minimum must be w-harmonic. Since the unique w-harmonic function satisfying the boundary condition is $\tilde{V}$, this gives the expression. Hence we can write

$$\mathcal{E}(\tilde{V}, w) = \inf \frac{1}{2} \sum_{x \in S} \sum_{y \in S} w(x,y)[f(y) - f(x)]^2,$$

where the infimum is over all functions f satisfying the boundary conditions. This formula may not look nicer than what we started with, but it does have a nice property. If we increase the weights on any edge, leaving the remaining weights the same, the value on the right-hand side clearly cannot get any smaller. In particular, if the weights (conductances) of an electrical network are increased, the effective resistance of the network decreases (i.e., does not increase). This fact is "obvious" from an electrical network perspective. Its mathematical proof does not use the electrical network intuition, but it was this intuition that told us that this fact had to be true!

Another interpretation of effective resistance is in terms of "escape probabilities". Let $\tilde{V}$ be as above. We have already seen the interpretation of $\tilde{V}(x)$ as the probability that the random walk starting at x visits some site in B_1 before visiting a. The current flowing through a is

$$\sum_{x \in S} w(a,x)\tilde{V}(x) = \frac{1}{V_0} = \frac{1}{\text{effective resistance}}.$$

Let p be the probability that the random walker starting at a visits B_1 before returning to a. Then

$$p = \sum_{x \in S} P(a,x)\tilde{V}(x) = \sum_{x \in S} \frac{w(a,x)}{d(x)}\tilde{V}(x),$$

or

$$d(x)p = \frac{1}{\text{effective resistance}}.$$

Consider random walk on the square lattice, and let R_n be the set of points (z_1, z_2) with $|z_1| + |z_2| = n$. Let q_n be the probability that a simple random walker starting at the origin reaches R_n before returning to the origin. We have already seen that a random walk is recurrent in two dimensions, and hence

$$\lim_{n \to \infty} q_n = 0.$$

We can think of simple random walk as being the random walk on the network with $w(x,y) = 1$ if x, y are nearest neighbors and $w(x,y) = 0$ otherwise. Note that by the formula above the effective resistance between 0 and R_n in this networks is $(4q_n)^{-1}$. Suppose some edges are removed from the lattice, i.e., for some x, y that are nearest neighbors we make $w(x,y) = 0$. Then the effective resistance must still be at least $(4q_n)^{-1}$. If we take a simple random walk on this grid with some edges removed, we can use the formula above (assuming the nontrivial case $d(0) \geq 1$) to show that the probability this new random walker reaches R_n before returning to the origin goes to 0 as $n \to \infty$. Therefore, random walk on this new grid is also recurrent. It is quite reasonable to believe that random walk on a sparser grid is more recurrent; the notion of effective resistance works very well as a tool for proving it.

We end this discussion by saying that a lot of what we have discussed holds equally well in continuous time and continuous space. As an example, consider a bounded open region $D \subset \mathbf{R}^d$ with (smooth) boundary ∂D. Suppose that voltages $V(x)$ are put on ∂D; for ease we will assume that $V : \partial D \to \mathbf{R}$ is a continuous function. The Laplacian in $\mathbf{R}^d$ is the operator

$$\Delta f(x) = \sum_{i=1}^{d} f_{ii}(x),$$

where f_{ii} denotes the second partial with respect to the ith component. The Dirichlet problem is to find a continuous function $f : D \cup \partial D \to \mathbf{R}$ such that

$$f(x) = V(x), \quad x \in \partial D,$$
$$\Delta f(x) = 0, \quad x \in D.$$

There is a unique solution to this problem, and a probabilistic solution can be given. Let B_t be a d-dimensional Brownian motion (a d-dimensional Brownian motion is a process that does independent one-dimensional Brownian motions in each component). Let $T = \inf\{t : B_t \in \partial D\}$ be the first time that the Brownian motion reaches the boundary. Then the unique solution to the Dirichlet problem is

$$f(x) = \mathbf{E}[V(B_t) \mid B_0 = x].$$

Lecture 10

Uniform Spanning Trees

A graph $G = (S, E)$ is called a *tree* if G is connected, and between any two vertices there exists only one self-avoiding path of edges. Equivalently, a connected graph is a tree if it contains no loops. We leave it as an exercise to show that any tree on S vertices has exactly $\#(S) - 1$ edges.

For a connected graph $G = (S, E)$, we will say that a subset of the edges $F \subset E$ forms a *spanning tree* if $G = (S, F)$ is a tree. Any connected graph has at least one spanning tree, but in general there are many spanning trees. Each spanning tree contains exactly $\#(S) - 1$ edges. One way to obtain a spanning tree is as follows: write down all the edges in order, $E = \{e_1, \ldots, e_n\}$; let $F_1 = \{e_1\}$. For $j = 2, \ldots, n$, define F_j recursively as follows: if $\{x_j\} \cup F_{j-1}$ contains a loop, let $F_j = F_{j-1}$; otherwise, let $F_j = \{x_j\} \cup F_{j-1}$. It is easy to check that F_n is a tree. In fact, if an infinite sequence of edges $e^1, e^2, \ldots$ is given (with repetitions, of course, since the total number of distinct edges is finite), then as long as each edge appears once in the sequence we can carry out the same procedure. Note that by the time we have seen all the edges in the sequence, we have our spanning tree.

Let $\mathcal{F} = \mathcal{F}(G)$ denote the set of spanning trees of the graph $G = (S, E)$. Given the graph, we can write down all the elements of $\mathcal{F}$ even if we see no particularly efficient way of doing it (if nothing else, we can list all $n-1$ element subsets of E and see which ones are trees). It is not easy to see immediately by looking at the graph how may spanning trees there are. A *uniform spanning tree* is a tree chosen from the uniform distribution on $\mathcal{F}$. We would like a way to sample from this distribution easily. We would also like a way to describe this distribution so that we answer questions such as: if $e_i, e_j \in E \ \ (i \neq j)$, is it true that

$$\mathbf{P}\{e_i \in F, e_j \in E\} \leq \mathbf{P}\{e_i \in F\}\mathbf{P}\{e_j \in F\},$$

or equivalently, that

$$\mathbf{P}\{e_j \in F \mid e_i \in F\} \leq \mathbf{P}\{e_j \in F\}? \tag{10.1}$$

The probabilities here are with respect to the uniform measure on $\mathcal{F}$. There is good intuitive reason to believe this inequality is true, since having $e_i \in F$ gives "less room" in F for e_j. However, it is not so clear if it is true in general.

We will use random walk on the tree to give an algorithm for sampling from the set of uniform spanning trees. Then we will use the relationship between random walk and electrical networks, to describe the above conditional probability in terms of resistance. This will allow us to conclude the inequality. The proof gives a nice combination of a problem in graph theory analyzed using random walk and using electrical networks.

How do we choose a spanning tree at random from the uniform distribution on all spanning trees? Consider a simple random walk on the graph. This is a simple random walk as defined in the previous lecture, where we take the weight function $w(x, y)$ to be 1 if x is adjacent to y (and 0 otherwise). Start at any vertex x; let $Y_0 = x$ and let Y_i be the vertex visited at time i; then choose our edges $X_1, X_2, X_3, \ldots$, where X_i is the edge connecting Y_{i-1} and Y_i. Eventually (with probability one) we will traverse all the edges of the graph. Hence we can derive a (random) spanning tree by performing the algorithm listed above for the random sequnce of edges $X_1, X_2, \ldots$. Pemantle [**P**] calls this algorithm the groundskeeper's algorithm. This algorithm is certainly easy to implement on a computer. The amazing thing is that the probability that a certain spanning tree is chosen in this algorithm is exactly the same for each spanning tree—this algorithm chooses a spanning tree from the uniform distribution!

We will use Markov chains to prove this fact. Recall that a simple random walk is a reversible Markov chain with invariant measure

$$m(y) = \frac{d(y)}{\sum_{x \in V} d(x)}.$$

Here $d(x)$ is the number of vertices adjacent to x. Let us run this chain until it reaches equilibrium; another way of thinking of this is to consider a doubly infinite sequence of random vertices

$$\ldots, Y_{-2}, Y_{-1}, Y_0, Y_1, Y_2, \ldots.$$

Here we choose Y_0 according to the invariant measure, and we choose $Y_1, Y_2, \ldots$ by running the walk "forwards" and $Y_{-1}, Y_{-2}, \ldots$ by running the walk "backwards". The reversibilty of the random walk on the graph tells us that the two directions look the same in the statistical sense (the forward transition probabilities are the same as the backwards transition probabilities, i.e., one always goes to a random adjacent vertex). This also gives us a doubly infinite sequence of edges

$$\ldots, X_{-2}, X_{-1}, X_0, X_1, X_2, \ldots,$$

where X_i is the edge connecting Y_{i-1} to Y_i.

Let Z_n be the random variable taking values in $V \times \mathcal{G}$,

$$Z_n = (Y_{-n}, T_{-n}).$$

In other words, Z_n records the position of the walker at time $-n$ and the directed tree that is formed from the random walker starting at that point.

What we will see is that Z_n is an irreducible Markov chain on $V \times \mathcal{F}$ whose transition matrix P is reversible, i.e., if $x, y \in V$ and $T^1, T^2 \in \mathcal{F}$, then

$$d(x)P[(x, T^1), (y, T^2)] = d(y)P[(y, T^2), (x, T^1)]. \tag{10.2}$$

Note that this implies that the equilibrium $\bar{m}$ for this chain distribution is proportional to $d(x)$. Hence if we start at any x and do the groundskeeper's algorithm, we choose a spanning tree from the uniform distribution on spanning trees.

To show that Z_n is a Markov chain we have to show that the transition probability of going from Z_n to Z_{n+1} depends only on Z_n, and not on any other information. Recall that Z_n is derived by doing the algorithm on the set of sites

$$Y_{-n}, Y_{-n+1}, \ldots, \tag{10.3}$$

while Z_{n+1} is obtained by considering the sites

$$Y_{-n-1}, Y_{-n}, \ldots \tag{10.4}$$

How do the trees formed from these sites differ? Imagine that when we traverse an edge, we put a little arrow down indicating which way we traversed that edge. Suppose that we knew that $Z_n = (x, T^1)$. Then we would know which way the arrows point. In the algorithm, whenever a site is visited for the first time, that edge is chosen. However, whenever a vertex is returned to, the edge is not chosen since either that edge has already been chosen, or adding that edge would produce a loop. Hence every vertex, except the initial vertex, has exactly one edge with the arrow pointing toward that vertex. This is the edge corresponding to the first visit to that vertex. Assume $Y_{-n} = x$ and $Y_{-n-1} = y$. The directed trees Z_n, Z_{n+1} differ in only one edge. The first tree has an edge pointing to vertex y; the second does not have this edge but does have the edge pointing from y to x. This tells us a method to get from Z_n to Z_{n+1}. Again, assume $Z_n = (x, T^1)$. Choose a vertex y uniformly from the vertices adjacent to x. Let T^2 be the directed graph which is the same as T^1 except that the edge going into y is removed and the directed edge from y to x is added. It is not difficult to see that this gives an irreducible Markov chain and that the transition matrix satisfies (10.2).

Now that we know that the groundskeeper algorithm started at any vertex produces a uniform spanning tree, we can use it to answer the question that we started with. First, we will give an electrical network expression for the probability that a certain edge is chosen in a spanning tree. Let $\{x, y\}$ be an edge and suppose we start our random walker at x. Let p be the probability that the edge $\{x, y\}$ will be included in the tree. Let q be the probability that the random walker starting at x visits y before returning to x. There are two ways to visit y before returning to x—either one can go immediately to y, with probability $1/d(x)$, or one can go to a different site initially and then visit y before returning to x. The latter case happens with probabilty $q-(1/d(x))$, by definition of q. In the former case, the edge $\{x, y\}$ is included in the tree, but in the latter case the edge $\{x, y\}$ is not included since a different edge going into y will have been chosen. If the random walker gets back to x without having visited y, then it just tries again. From this we see that

$$p = \frac{1}{d(x)} + (1-q)p,$$

or

$$p = \frac{1}{qd(x)}.$$

Let $V(z)$ be the function that is harmonic (with respect to the weight described above) for all $z \neq x, y$, and has the boundary values $V(x) = 0, V(y) = 1$. Then we saw in the last lecture that

$$q = \sum_{z \sim x} \frac{1}{d(x)} V(z).$$

Hence, recalling that $V(y) = 1, V(x) = 0$, we can write

$$p = \frac{V(y) - V(x)}{\sum_{z \sim x}[V(z) - V(x)]}.$$

In other words, p represents the fraction of the current going through the edge x, y if we place voltages 0 at x and 1 at y. We have written the probability of an edge being in a uniform spanning tree in terms of the current flow in an electrical network. Recall that the effective resistance between x and y in the network is given by

$$\left(\sum_{z \sim x}[V(z) - V(x)]\right)^{-1}.$$

Hence we see that p is the effective resistance between x and y in the network.

Now suppose we have a different edge $\{z, w\}$, and we want to choose a spanning tree uniformly from all spanning trees that include $\{z, w\}$. In other words, we want the conditional distribution on uniform spanning trees, given that this edge is in the tree. It is not difficult to see that this is the same thing as choosing a uniform spanning tree from the graph which identifies the two vertices z and w

(this new graph may have pairs of vertices with two edges connecting them, but this will not affect the analysis). From an electrical network viewpoint, we change our network by putting a zero resistance between $\{z, w\}$. By what we know about electrical networks, decreasing the resistance on one edge decreases the effective resistance. Hence the probability that $\{x, y\}$ is chosen, given that $\{z, w\}$ is chosen, is less than or equal to the unconditional probability p.

Lecture 11

Random Walk Simulations

In the final three lectures we will consider Monte Carlo simulations. Simulations are a valuable tool for understanding probabilistic models that are difficult to understand analytically. We will first consider random walk simulations in dimension $d = 1$. In Lecture 1 we learned that $\mathbf{E}[S_n^2] = n$, i.e., that the square of the expected value of the position of the simple random walk (started at zero) after n steps is equal to n. It is quite easy to simulate the simple random walk and check that $\mathbf{E}[S_n^2] = n$. One starts a program by setting $S_0 = 0$. Then one calls on a random number generator to determine whether the first jump will be to the left or right.

Most computer systems have a random number generator whose output is a sequence of pseudo random numbers that represent independent samples from the uniform (0,1) distribution or are independent samples that are uniform on the integers $0, 1, \ldots, n$ for some specified number n. They are called pseudo random because they are not actually random; however, for practical purposes one assumes that they are random. The algorithm used to generate the pseudo random numbers is often of the form

$$x_{n+1} = (ax_n + b) \text{ modulo } c.$$

a, b, and c are usually very large (relatively co-prime) numbers. Once a seed (x_0) has been chosen, then $x_1, x_2, \ldots$ can be generated. In particular, if we use the same seed, we will generate the same sequence of random numbers (which can be useful if we wish to compare two different algorithms or if we need to debug a program). In writing a Maple procedure, we can use *rand*(2), which outputs 0 with probability 1/2 and outputs 1 with probability 1/2. Thus, for our random walk simulation, we can set $S_1 = 1$ if $rand(2) = 0$ and $S_1 = -1$ if rand(2)=1. If we are programming in the C language, we can use the random number generator *drand*48( ) which produces a number that is uniformly distributed on [0,1). Thus if $drand48(\ ) < 0.5$ we can let $S_1 = 1$ and if $drand48(\ ) \geq 0.5$, we let $S_1 = -1$. We can continue to call on the random number generator to produce S_2, S_3, ... After n calls to the random number generator, we would have a value for S_n. This produces one *realization* of S_n. To estimate $\mathbf{E}[S_n^2]$ we would need to go through the above procedure a large number of times (perhaps 10,000 realizations) and average over the values that we obtain for S_n.

Of course, we can check many other simple random walk facts. In lecture 1 we also learned that

$$\mathbf{P}\{S_{2n} = 0\} \sim \frac{1}{\sqrt{\pi n}}, \qquad \mathbf{E}(R_n) \sim \frac{2n^{1/2}}{\sqrt{\pi}},$$

where R_n is the number of visits to the origin up through time $2n$. Since these are asymptotic properties, they are not quite as straightforward to check. We thus need some tools for estimating asymptotic parameters.

Suppose that one has some data points $(x_1, y_1), \ldots, (x_n, y_n)$ and one wants to fit these data points to some nice curve. Let us first consider the problem of finding the best line

$$y = ax + b$$

to fit the data. We have to make a decision as to what criteria we will use for determining the "best" fit. The most common criterion is the least squares fit: find the a, b such that the sum of the squared deviations

$$L(a,b) \doteq \sum_{j=1}^{n} [(ax_j + b) - y_j]^2$$

is as small as possible. This is a natural criterion, although there are a number of other natural criteria we could use. One of the reasons this one is chosen is that it is not very difficult to find the a and b which minimize $L(a,b)$. From multivariable calculus, we know that at the (a,b) at which $L(a,b)$ takes on a minimum,

$$\frac{\partial}{\partial a} L = 0, \qquad \frac{\partial}{\partial b} L = 0.$$

Hence, at the optimal (a,b),

$$\sum_{j=1}^{n} 2x_j[ax_j + b - y_j] = 0,$$

$$\sum_{j=1}^{n} 2[ax_j + b - y_j] = 0.$$

After dividing by $2n$, we can write these equations as

$$a\langle x^2\rangle + b\langle x\rangle - \langle xy\rangle = 0,$$

$$a\langle x\rangle + b - \langle y\rangle = 0,$$

where

$$\langle x\rangle = \frac{1}{n}(x_1 + \cdots + x_n),$$

$$\langle x^2\rangle = \frac{1}{n}(x_1^2 + \cdots + x_n^2),$$

$$\langle y\rangle = \frac{1}{n}(y_1 + \cdots + y_n),$$

$$\langle xy\rangle = \frac{1}{n}(x_1y_1 + \cdots + x_ny_n).$$

By solving the equations, we get

$$a = \frac{\langle xy\rangle - \langle x\rangle\langle y\rangle}{\langle x^2\rangle - \langle x\rangle\langle x\rangle},$$

$$b = \langle y\rangle - a\langle x\rangle.$$

Often, data looks as if it fits not a line, but rather some more complicated curve. In the work we are doing we will encounter data that more closely approximate functions f of the form

$$f(x) = cx^{\alpha} \tag{11.1}$$

or

$$f(x) = C(\ln x)^{\beta}. \tag{11.2}$$

If we have data $(x_1, y_1), \dots, (x_n, y_n)$ and want to fit it to a curve of type (11.1), we can consider instead the data

$$(\ln x_1, \ln y_1), \dots, (\ln x_n, \ln y_n)$$

and fit it to a straight line

$$\ln y = \ln c + \alpha \ln x.$$

If we want to fit to an equation of type (11.2) with $\beta = 1$ we can consider the data $(\ln x_1, y_1), \dots, (\ln x_n, y_n)$, and see whether there is a nice fit with a line. If we want to fit the data with other β we can try fitting the data $(\ln x_1, y_1), \dots, (\ln x_n, y_n)$ to an equation of type (11.1). This eventually requires looking at the data $(\ln \ln x_1, \ln y_1), \dots, (\ln \ln x_n, \ln y_n)$. Theoretically this sounds OK, but practically it is often not very useful because $\ln \ln$ goes to infinity very slowly.

We can now proceed to estimate $\mathbf{P}\{S_{2n} = 0\}$ and $\mathbf{E}(R_n)$. We generate the sequence $S_0, S_1, \dots$ as before, and every time that $S_{2j} = 0$ (for some integer $j \geq 0$) we add one to a counter (R, say) that stores the number of times that the walk returns to zero. Then, for every e.g., $m = 100 \times k$, $k = 1, 2, \dots, 100$, we store the current value of R (equals R_m), and whether $S_{2m} = 0$. We repeat this procedure 10,000 times (say) to find estimates for $\mathbf{E}(R_m)$ (average over all the values for R_m) and $\mathbf{P}\{S_{2m} = 0\}$ (find the proportion of times that $S_{2m} = 0$), for $m = 100, 200, \dots, 10,000$. Now that we have 100 pieces of data for $\mathbf{E}(R_m)$ and $\mathbf{P}\{S_{2m} = 0\}$, we can use log plots and least squares to estimate their asymptotic behavior. Note that the numbers that we choose for the maximum walk length and the number of simulations will depend on the language and computer that you are using. For instance, a program written in C will generally run significantly faster than a Maple procedure, and thus one can run many more simulations in C than in Maple (in the same amount of time). We should start with quite small numbers to ensure that our program will run in a reasonable amount of time, and then experiment with larger values. Of course, in general the larger the number of simulations, the more accurate our estimate. In order to improve our estimates, we may wish to ignore the earliest values of the quantities that we are trying to estimate (e.g., only consider $\mathbf{P}\{S_{2m} = 0\}$ for values $m = 2,000, 2,100, \ldots, 10,000$). This is reasonable since we are trying to find the aysmptotic behavior as the number of steps goes to infinity, and the first few values of the quantity can be significantly different from the values predicted from the asymptotics.

Now we consider the d-dimensional simple random walk. For instance, in two dimensions, if we are using Maple we use $rand(4)$, and if $rand(4) = 0$, we set $S_1 = (1, 0)$; if $rand(4) = 1$, set $S_1 = (-1, 0)$; if $rand(4) = 2$, set $S_1 = (0, 1)$; and if $rand(4) = 3$, set $S_1 = (0, -1)$. If we are writing a C program, we set $S_1 = (1, 0)$ if $drand48() < 0.25$, etc. Again, we can use the simulations to verify that $\mathbf{E}[S_n^2] = n$ and to estimate $\mathbf{E}(R_m)$ and $\mathbf{P}\{S_{2m} = 0\}$.

In Lecture 2 we discussed the probability that random walks do not intersect. In particular we noted that q_n decays like $n^{-\zeta}$, where $\zeta = \zeta(d)$, and q_n is the probability that the paths of two random walkers S and W have not intersected by time n. We now would like to discuss how we can estimate ζ using simulations. This program requires significant use of memory and a large number of simulations, so we need to use a language such as C. There are many different ways to program these simulations—here we just give one approach for the two-dimensional case. Suppose we wished to consider walks of length N. Then we would create a two-dimensional array (called $Grid$, say) of size $2N + 1 \times 2N + 1$. The counter $Grid$ would have one of three values: 0 if none of the walks have been to the corresponding point on the grid (note that we "center" the grid, so $Grid[N][N]$ represents the origin), 1 if walker S has been to the corresponding point, and 2 if walker W has been there. We simulate the walks of both walkers S and W simultaneously, storing their paths in $Grid$. After each step, we test whether there has been an intersection. If we find an intersection, we stop this simulation and move on to a second simulation. If we find no intersection after N steps, we chalk up a win. We wish to simulate many different realizations of the paths of walkers S and W. Instead of initializing the *Grid* array each time (which is time consuming; often the walkers only get to make a few steps), we also have another two-dimensional array (called *Number*, say) of the same size as *Grid* that stores the number of the simulation for which the corresponding *Grid* point was last visited by a walker. For example, suppose we are running our tenth simulation, and walker S steps to the point corresponding to $Grid[N+7][N-17]$. If the value of $Grid[N+7][N-17]$ is equal to 1, then there is no intersection, and we update $Number[N+7][N-17]$ to be equal to 10. If the value of $Grid[N+7][N-17]$ is equal to 2, then walker W was the last walker at this point, and so we need to check $Number[N+7][N-17]$. If $Number[N+7][N-17]$ is equal to 8, then this means that the last time a walker visited $Grid[N+7][N-17]$ was on the eighth simulation; thus there is no intersection, and we update $Number[N+7][N-17]$ to be 10 and $Grid[N+7][N-17]$ to be equal to 1. If $Number[N+7][N-17]$ is equal to 10, then we do have an intersection, and we move on to the next simulation. After a large number of simulations, we use the parameter-estimation methods already discussed to estimate ζ. If memory becomes an issue, we can use a combination of static memory (grid of size $4\sqrt{N} \times 4\sqrt{N}$ say) and dynamic memory (to catch the rare walkers that get to "fall off" the grid). This procedure can be generalized to more than two dimensions, although we soon run into memory problems in higher dimensions.

In the two dimensional case, we ran 10,000,000 pairs of walks of length 5,000, and this led to estimates for $\zeta(2)$ that were consistently in the interval $[.621, .623]$. This value is slightly lower than the conjectured value of .625, but it is possible that we have not taken sufficiently long walks to reach the asymptotic regime. In the three dimensional case, we ran 10,000,000 pairs of walks of length 2,500, and this led to estimates for $\zeta(3)$ that were consistently in the interval $[.288, .289]$.

Lecture 12

Other Simulations

In Lecture 4 we discussed Brownian motion, B_t, and we learned that $B_t - B_s$ has a normal distribution with mean zero and variance $t - s$, N(0, $t - s$). To simulate a Brownian motion path we generally discretize time, e.g., we could plot a point on the path every 0.1 unit of time. Suppose we find a way to find random samples from the standard normal distribution, N(0, 1). If we multiply the random sample by $\sqrt{0.1}$, we get a sample from N(0, 0.1). This would then be our realization of $B_{0.1}$. We would then take another sample from N(0, 0.1) and add this value to $B_{0.1}$ to get $B_{0.2}$ and so on. In this way we can plot a realization of a Brownian motion path over 10 units of time by plotting 100 points $B_{0.1}, B_{0.2}, \dots, B_{10}$. If we were just interested in the value of B_{10} we would just take a random sample from N(0, 1) and multiply it by $\sqrt{10}$. In this way, we could find a large number of samples of B_{10}, and check that $\mathbf{E}[B_{10}] = 0$ and $\mathbf{E}[B_{10}^2] = 10$. Of course, since we are just simulating normal random variables, we would just be checking that our process for simulating normal random variables was actually generating numbers that had the correct mean and variance. But how do we simulate a normal random variable when we just have a random number generator that generates uniform random numbers between 0 and 1 at our disposal? (Maple and other CASs do have normal random variable generators, but if we are programming in C we need to write our own code.) We will now discuss one method of simulating normal random variables. Along the way, we will learn two general techniques for simulating continuous random variables using a random number generator.

The first technique that we will discuss is the *inverse transformation method.* This method is based on the following fact. Suppose U is a uniform random variable on $(0, 1)$, F is any continuous (cumulative) distribution function, and $Y = F^{-1}(U)$; then Y has distribution function F. To see why this is true, let F_Y be the distribution function of Y. Then for $a \in \mathbf{R}$, we have

$$\begin{aligned} F_Y(a) &= \mathbf{P}\{Y \leq a\} \\ &= \mathbf{P}\{F^{-1}(U) \leq a\} \\ &= \mathbf{P}\{U \leq F(a)\} \\ &= F(a). \end{aligned}$$

(Why can we go from the second line to the third line?) For an example, we will see how to simulate an exponential random variable with parameter λ. Let

$$F(x) = 1 - e^{-\lambda x}, \quad x \geq 0.$$

Note that F is the distribution function of an exponential random variable with parameter λ. Then $F^{-1}(x) = -\ln(1-x)/\lambda$. Let $Y = -\ln(1-U)/\lambda$. Then the fact that we discussed above tells us that Y is exponentially distributed with parameter λ. Since $1 - U$ is also uniform on $(0, 1)$, we can let $Y = -\ln(U)/\lambda$. So, to find a sample y from the exponential distribution with parameter λ, we take a sample u from U, and let $y = -\ln(u)/\lambda$.

To use the inverse transformation method we need to find an analytic expression for F^{-1}; for many distributions, such as the normal distribution, this is not possible. However, just as we used samples from a uniform to simulate an exponential random variable, we can use samples from an exponential to simulate a normal random variable. This second technique is called the *rejection method.* Suppose we wish to simulate a random variable X with (probability) density function f, and we know how to simulate a random variable Y with density function g (with the support of f equal to the support of g).

Also suppose that there exists a constant c such that

$$\frac{f(y)}{g(y)} \leq c \quad \text{for all } y \in \text{supp}(g).$$

To find a sample x from the distribution of X, the algorithm goes as follows.

1. Simulate $y \in Y$ and $u \in U$ (where U is uniform on $(0,1)$ as before).
2. If $u \leq f(y)/(cg(y))$, let $x = y$. Otherwise return to 1.

To see that x does correspond to a sample from the distribution of X, note that

$$\begin{aligned} \mathbf{P}\{X \leq x\} &= \mathbf{P}\left\{Y \leq x \,\middle|\, U \leq \frac{f(Y)}{cg(Y)}\right\} \\ &= \mathbf{P}\left\{Y \leq x; U \leq \frac{f(Y)}{cg(Y)}\right\} /K, \end{aligned}$$

where $K = \mathbf{P}\{U \leq f(Y)/(cg(Y))\}$. By independence, the joint density function of Y and U is

$$f(y,u) = g(y), \quad 0 < u < 1,$$

which implies

$$\begin{aligned} \mathbf{P}\{X \leq x\} &= \frac{1}{K}\int_{-\infty}^{x}\left(\int_0^{f(y)/(cg(y))} du\right) g(y)\,dy \\ &= \frac{1}{cK}\int_{-\infty}^{x} f(y)\,dy. \end{aligned}$$

If we let $x \to \infty$, we get

$$1 = \frac{1}{cK}\int_{-\infty}^{\infty} f(y)\,dy = \frac{1}{cK}.$$

Therefore we have

$$\mathbf{P}\{X \leq x\} = \int_{-\infty}^{x} f(y)\,dy.$$

As an example, if we wished to simulate a standard normal random variable Z, we would first let $X = |Z|$. Then

$$f_X(x) = f(x) = \frac{2}{\sqrt{2\pi}}\, e^{-x^2/2} \quad \text{for } 0 < x < \infty.$$

We know how to simulate an exponential random variable Y with parameter 1. Then

$$\begin{aligned} g(y) = e^{-y} \Rightarrow \frac{f(y)}{g(y)} &= \frac{2}{\sqrt{2\pi}}\, e^{-y^2+y} \\ &= \sqrt{2e/\pi}\, e^{-(y-1)^2/2} \\ &\leq \sqrt{2e/\pi}, \end{aligned}$$

where we completed the square on the third line. Thus a good candidate for c is $c = \sqrt{2e/\pi}$. This gives

$$\frac{f(y)}{cg(y)} = e^{-(y-1)^2/2}.$$

The algorithm for simulating a standard normal random variable goes as follows:

1. Generate a sample, y, from the exponential distribution with parameter 1, and an independent sample, u_1, from the uniform $(0,1)$ distribution.

2. If $u_1 \leq \exp[-(y-1)^2/2]$, let $x = y$. Otherwise, return to 1.

3. Generate another (independent) sample, u_2, from the uniform $(0,1)$ distribution, and let

$$z = \begin{cases} x & \text{if } u_2 < 0.5, \\ -x & \text{if } u_2 \geq 0.5. \end{cases}$$

Then z will be a sample from the standard normal distribution. If we wished to simulate a normal random variable $\tilde{z}$ with mean μ and variance σ^2, we would just simulate a sample z from the standard normal distribution, and let $\tilde{z} = \mu + \sigma z$. If we are using simulated normal r.v.'s in order to find an accurate estimate of a certain quantity, we can improve our accuracy by using variance reduction techniques. We shall not discuss these techniques here.

In Lecture 5 we discussed random permutations. In particular, in Example 4, we discussed the random k-transposition, which gives an algorithm for creating a random permutation on a computer. We will now provide some details for this algorithm.

We need to be able to simulate a number that is equally likely to take on any of the values $1, 2, \ldots, k$. One way to do this is to take u a sample from the uniform (0,1) distribution and let $m = [uk] + 1$, where $[x]$ is the integer part of x. Then the algorithm goes as follows:

1. Let $x_1, \ldots, x_N$ be any permutation of $1, 2, \ldots, N$. (For simplicity, one could take $x_i = i$.)
2. Let $j = 2$.
3. Let $m = [ju] + 1$ (where u is a sample from uniform (0,1)).
4. Interchange the values of x_j and x_m.
5. If $j < N$, increase the value of j by 1 and return to 3.
6. $x_1, \ldots, x_N$ is the desired random permutation.

We can generalize this algorithm to create a random permutation of any list of real numbers (i.e., where $x_i = r_i \ \forall i$). These algorithms are also used in sorting data. Often, it is more efficient to create a random permutation of one's data and then sort rather than attempting to sort the data as it already is.

In Lecture 8 we discussed Markov Chain Monte Carlo techniques. Here we wish to give some details regarding the algorithm for estimating the number of matrices of 0s and 1s, with neighboring 1s forbidden, of size 50×50, i.e., the cardinality of T_{50}. Let M_k be the set of matrices M in T_{50} for which the first $5k$ slots are zero where we define the (i, j)th matrix entry to be the $(50 \times (i-1) + j)$th slot. For example, matrices in M_{10} have zeros in the entire top row, matrices in M_{20} have zeros in the entire two top rows, etc. Note that our choice of 5 (in $5k$) is somewhat arbitrary. We have chosen 5 since it is a divisor of 50, and $\mathbf{P}(M \in M_k | M \in M_{k-1}) \approx 0.2$ (see below).

The probability that $M \in T_{50}$ is equal to the zero matrix (i.e., $M(i,j) = 0 \ \forall i, j$) is

$$\mathbf{P}(M = 0) = \prod_{k=1}^{500} \mathbf{P}(M \in M_k | M \in M_{k-1}),$$

since

$$\mathbf{P}(M \in M_k \,|M \in M_{k-1}) = \frac{\mathbf{P}(M \in M_k)}{\mathbf{P}(M \in M_{k-1})}$$

and $\mathbf{P}(M \in M_0) = 1$, where $\mathbf{P}(\,\cdot\,)$ is with respect to the uniform measure on T_{50}. We can estimate $\mathbf{P}(M \in M_k \,|M \in M_{k-1})$ by simulation. We start with a matrix of all zeros (say). Then we follow the procedure to generate the next matrix in the chain as outlined in Lecture 8. However, we must keep all zeros in the first $5(k-1)$ slots. Thus we choose an entry in the remaining $2500 - 5(k-1)$ slots at random. If the entry is a 1, change the entry to a 0; if the entry is a 0, change it to a 1 unless this would result in a matrix that is not in T_{50} — if the change is not permitted, then make no change to the matrix; output the matrix we now have. To estimate the required probability, we sample from the chain after every 2,500 iterations (say). (If we do 2,500 iterations, then every matrix entry has a reasonable chance of being changed.) We generate 10,000 samples (which means we iterate the chain $10,000 \times 2,500$ times) and find the proportion of times that the sample will have all zeros in the five slots $5k-4$ through $5k$.

Once we have found an estimate for $\mathbf{P}(M = 0)$, we can invert this number to get an estimate of the total number of matrices in T_{50}. (Why?) If we let $F(N^2)$ be equal to the number of matrices in T_N, and

$$\eta = \lim_{N\to\infty} F(N^2)^{1/N^2},$$

then it can be proven (see [**CW**]) that $1.50304\ldots \leq \eta \leq 1.50351\ldots$. Our simulations in the case of $N = 50$ led to an estimate of 1.50719 for $F(50^2)^{1/50^2}$.

Lecture 13

Simulations in Finance

Here we will discuss Monte Carlo simulations in relation to finance. A model for the behavior of stock prices that is often used is geometric Brownian motion (see [**H**] for an introduction to mathematical finance). Geometric Brownian motion can be described by the stochastic differential equation

$$dS = \mu S dt + \sigma S dB, \tag{13.1}$$

where S is the stock price, t is time, B is Brownian motion, μ is the (percentage) drift, and σ is the (percentage) volatility. In this discussion we will assume that μ and σ are constants. To fully understand this equation, we would need to discuss stochastic integration. Rather than do that, we will describe how to simulate such a process. By discussing the simulation, we hope also to give some idea of what the equation means. For comparison purposes, suppose we had a deterministic differential equation such as

$$dy = (y + t)dt, \tag{13.2}$$

and suppose that we did not know how to explicitly solve this equation. (As we note below, this equation does have a simple solution, but for now we will assume that we do not know the solution.) If we were given an initial condition, such as $y(0) = 1$, and we wanted to estimate $y(T)$, then we could discretize the differential equation to give

$$\begin{aligned} \Delta \tilde{y} &= (\tilde{y} + t)\Delta t \\ \Rightarrow \tilde{y}(t_{j+1}) &= \tilde{y}(t_j) + (\tilde{y}(t_j) + t_j)(t_{j+1} - t_j). \end{aligned}$$

By making $t_{j+1} - t_j$ sufficiently small, $\tilde{y}(T) \approx y(T)$. This is known as Euler's method. In a similar way, we can discretize the stochastic differential equation (13.1) and get

$$\Delta \tilde{S} = \mu \tilde{S} \Delta t + \sigma \tilde{S} \Delta B \tag{13.3}$$

$$\begin{aligned} \Rightarrow \tilde{S}(t_{j+1}) = \tilde{S}(t_j) + \mu \tilde{S}(t_j)(t_{j+1} - t_j) \\ + \sigma \tilde{S}(t_j)\sqrt{t_{j+1} - t_j} X, \end{aligned} \tag{13.4}$$

where X is the standard normal random variable. Therefore we can see that if Δt is small, one day say, where time is measured in years, then the change in the stock over one day is approximately normal with mean $\mu S(t_0)/365$ and standard deviation $\sigma S(t_0)/\sqrt{365}$. This is one way of seeing why μ is considered the (yearly percentage) drift and σ the (yearly percentage) volatility. (Since in general the change in the market between Friday and Monday is similar to the change between two week days, people often think of one trading day as $1/N$ years, where N is the number of trading days in one year (approximately 250).) We can simulate a path for the stock price over a time period $[0, T]$ by dividing the period of time into n intervals $[0, T/n], \ldots,$ $[(n-1)T/n, T]$, and using equation (13.4) and our algorithm to simulate a normal random variable to find $S(T/n)$, etc. We will refer to this method as the first method (for simulating a stock price path).

The deterministic equation (13.2) actually does have an analytic solution, $y(t) = -t - 1 + c\exp(t)$. Thus, we do not need Euler's method to find $y(T)$. Similarly, for our stochastic differential equation (13.1), we would like to find an "analytic" solution. It can be shown, using a result called Ito's lemma, that the analytic solution of equation (13.1) is given by

$$S(t_2) = S(t_1) \exp\left[\left(\mu - \frac{\sigma^2}{2}\right)(t_2 - t_1) + \sigma\sqrt{t_2 - t_1} X\right], \tag{13.5}$$

where $t_1 < t_2$ and X is standard normal. We can use this equation to simulate a stock price path in the same manner as the first method. If we want to use Monte Carlo simulation to estimate $\mathbf{E}[f(S(T))]$ for some function f, we will always use this second method since it does not involve any "approximation" and thus will give more accurate answers than the first method (see Exercises 13-1 and 13-2).

Referring to equation (13.5), suppose that we know $S(t_1)$. Then, taking natural logs, we get

$$\ln\{S(t_2)\} = \ln\{S(t_1)\} + \left(\mu - \frac{\sigma^2}{2}\right)(t_2 - t_1) + \sigma\sqrt{t_2 - t_1}X. \tag{13.6}$$

This shows that $\ln\{S(t_2)\}$ is normally distributed with mean

$$\ln\{S(t_1)\} + \left(\mu - \frac{\sigma^2}{2}\right)(t_2 - t_1)$$

and variance $\sigma^2(t_2 - t_1)$. Thus $S(t_2)$ is lognormally distributed, and from the properties of the lognormal distribution we have $\mathbf{E}[S(t_2)] = S(t_1)\exp\{\mu(t_2 - t_1)\}$.

A European call option on a stock S with strike K and expiration date T gives the holder the right to buy the stock S at time T for price K. Of course, if the stock price is less than K at time T, then the holder of the option will not exercise the option. If the stock price is greater than K at time T then the holder will exercise the option and purchase the stock for price K. The stock can be sold on the market for price $S(T)$, and so the cash profit is $S(T) - K$. Thus there is a payout at time T of $\max(S(T) - K, 0)$. We make the assumptions that we can buy stocks on the market for the same price that we can sell them for, that there are no transaction costs or taxes, that the interest rate, r, is constant over time, that one cannot make a sure profit (i.e., a profit without any risk), and a few other minor assumptions. Then Black-Scholes pricing theory tells us that the value c of the option at time t is

$$c = e^{-r(T-t)}\mathbf{E}[\max(S(T) - K, 0)], \tag{13.7}$$

where the expectation is taken with $\mu = r$ (this is referred to as the risk-neutral expectation; we are setting the drift of the stock to be equal to the interest rate). The price of the option is the (risk-neutral) average payout at time T discounted to time t. If we know $S(t)$, K, r, σ, and T, then we can simulate (with $\mu = r$) the price at time T using equation (13.5) (note that we do not need to simulate the entire stock path, we just need its price at time T) and calculate $\max(S(T) - K, 0)$.

We repeat this a large number of times, find the arithmetic average of $\max(S(T) - K, 0)$, and multiply by $e^{-r(T-t)}$ to get our estimate for c. There does exist a solution in terms of the standard normal distribution function to (13.7), so Monte Carlo simulation would not be very useful in this case. However, for more complicated options that depend on the path of the stock, such as Asian options, when the payoff is $\max(S(T) - S_{\mathrm{AVE}}, 0)!$ (where S_{AVE} is the average price of the stock over a certain period of time), and corridor options, when the payoff depends on the length of time that the stock price spends between two values K_1 and K_2, it is either impossible or very difficult to find an exact pricing formula, and thus Monte Carlo simulation is very useful.

Problems

1-1. Show that for every $s > -1$ the limit

$$L = \lim_{n\to\infty} n^{-s} \prod_{j=1}^{n} \left[1 + \frac{s}{j}\right]$$

exists and is positive.

1-2. Let

$$p_n = P\{S_{2n} = 0\} = 2^{-2n}\binom{2n}{n}.$$

Show that

$$p_n = \left[1 - \frac{1}{2n}\right] p_{n-1},$$

and hence

$$p_n = \prod_{j=1}^{n} \left[1 - \frac{1}{2j}\right].$$

Use this to show that the limit

$$\lim_{n\to\infty} n^{1/2} p_n$$

exists and is positive.

1-3. Let L be the undetermined constant in Stirling's formula,

$$L = \lim_{n\to\infty} \frac{n!}{n^n e^{-n}\sqrt{n}}.$$

Use Stirling's formula to give the approximation

$$\mathbf{P}\{S_{2n} = 2a\sqrt{n}\} \sim \frac{\sqrt{2}}{L\sqrt{n}} e^{-a^2},$$

assuming that $a\sqrt{n}$ is an integer.

1-4. Using Problem 3, derive the approximation

$$\mathbf{P}\{-2a\sqrt{n} \le S_{2n} \le 2a\sqrt{n}\} \sim \frac{\sqrt{2}}{L}\int_{-a}^{a} e^{-x^2}\,dx.$$

1-5. Recall Chebyshev's inequality: for any random variable Y and any $\delta > 0$,

$$\mathbf{P}\{|Y - \mathbf{E}(Y)| \ge \delta\} \le \frac{\mathrm{Var}(Y)}{\delta^2}.$$

Use Chebyshev's inequality and Problem 4 to conclude that

$$\frac{\sqrt{2}}{L}\int_{-\infty}^{\infty} e^{-x^2}\,dx = 1.$$

1-6. Use Problem 5 to conclude that $L = \sqrt{2\pi}$.

The purpose of the remaining problems is to investigate

$$q_n = \mathbf{P}\{S_1 > 0, S_2 > 0, \dots, S_{2n} > 0\}.$$

We already discussed the fact that the random walker will return to zero eventually. This implies that $q_n \to 0$. We will see how fast the convergence is.

1-7. If a, b, n are integers, let $K(a,b,n)$ be the number of random walk paths of n steps starting at a and ending at b. Note that $K(a,b,n) = K(a-i, b-i, n)$ for any integer i. Let $\tilde{K}(a,b,n)$ be the number of such paths such that the random walker crosses the origin in one of the n steps. Suppose $b > 0$. Show that

$$\tilde{K}(1,b,n) = \tilde{K}(-1,b,n) = K(-1,b,n).$$

1-8. Show that

$$\begin{aligned} q_n &= 2^{-2n}\sum_{b=1}^{\infty}[K(1,2b,2n-1)-\tilde{K}(1,b,2n-1)] \\ &= 2^{-2n}\sum_{b=1}^{\infty}[K(1,2b,2n-1)-K(-1,2b,2n-1)] \\ &= 2^{-2n}K(1,2,2n-1) \\ &= 2^{-2n}K(1,0,2n-1) \\ &= \frac{1}{2}\mathbf{P}\{S_{2n}=0\} \end{aligned}$$

1-9. Suppose a random walker starts at the origin. Consider the number of steps T until the walker returns to the origin for the first time. We have already seen that $T < \infty$ (with probability one). What is $\mathbf{E}(T)$?

1-10. Let T be a positive integer random variable with

$$\mathbf{P}\{T=n\} = cn^{-2}, \quad n = 1, 2, \ldots,$$

where

$$c = \left[\sum_{n=1}^{\infty}\frac{1}{n^2}\right]^{-1} = \frac{6}{\pi^2}.$$

What is $\mathbf{E}(T)$?

2-1. Suppose there are two indepedent one-dimensional random walkers starting at 0 and 2 respectively. Can we be sure that there is some time in the future that both walkers are at the same point?

How about the case of k independent walkers starting at $0, 2, \ldots,$ $(k-2)$? Does there exist some time when all k walkers are at the same point?

How about two two-dimensional walkers, one starting at the origin and one starting at $(1,1)$?

2-2. Let $P(d)$ be the probability that a simple random walker in d dimensions ever returns to the origin.

(a) Show that

$$\lim_{d\to\infty} P(d) = 0.$$

(b) Find

$$\lim_{d\to\infty} \frac{P(d)}{d}.$$

2-3. Consider a d-dimensional random walk and let x be a point in $\mathbf{Z}^d$ other than the origin. Let $Z = Z_x$ be the number of times that a random walker (starting at the origin) visits the point x before returning to the origin for the first time. Find $\mathbf{E}(Z)$. (Hint: you may wish to consider $\mathbf{E}[Z \mid Z > 0]$.)

2-4. For $d \geq 3$, let L be the size of the "loop" at the origin, i.e., L is the largest time n such that $S_{2n} = 0$. For which dimensions d is $\mathbf{E}[L] < \infty$?

2-5. Suppose that we have an infinite number of random walkers in one dimension, with exactly one random walker starting at each integer. Let Y_n be the number of these walkers who are at the origin at time n. Show that

$$0 < \lim_{n\to\infty} \mathbf{P}\{Y_n = 0\} < 1.$$

For any positive integer k, find

$$\lim_{n\to\infty} \mathbf{P}\{Y_n = k\}.$$

2-6. A set $A \subset \mathbf{Z}^d$ is called recurrent for simple random walk if a random walker is guaranteed to visit A an infinite number of times. Note that for $d = 1, 2$ all nonempty sets are recurrent, while for $d \geq 3$, recurrent sets have to be infinite. For which d is the line

$$L = \{(x^1, \ldots, x^d) : x^2 = x^3 \cdots = x^d\}$$

a recurrent set?

2-7. Consider d-dimensional random walk ($d \geq 3$). For any x, let $p(x)$ be the probability that a random walker starting at the origin ever visits the point x. What does $p(x)$ look like for large x? In particular, can you find which number s has the property that

$$0 < \lim_{|x| \to \infty} |x|^s p(x) < \infty?$$

3-1. Suppose a random walker starts at the origin in two dimensions. At each step the random walker looks at the four nearest neighbors and sees which one the walker has not visited. The walker then chooses randomly among these choices (after the first step there will be at most three possible choices). If there are no new places to move, the walker stops. Is it true that the walker always stops, or is it possible that the walker will be able to continue forever?

3-2. How many self-avoiding walks of length 10 are there in two dimensions that start at the origin and end at the point $(5,5)$?

3-3. How many SAWs of length 4 are there in two dimensions? What is the probability that a (simple) random walker visits five distinct sites (including the origin) in its first four steps?

3-4. Let K_n be the number of simple random walk paths in $\mathbf{Z}^2$ of length $2n$ that start at the origin and end at the origin. What is

$$\lim_{n \to \infty} K_n^{1/2n}?$$

Let $\tilde{K}_n$ be the number of "self-avoiding polygons" of length $2n$, i.e., sequences of points $(\omega_0, \ldots, \omega_{2n})$ with $\omega_0 = \omega_{2n} = 0$; $|\omega_i - \omega_{i-1}| = 1, i = 1, \ldots, 2n$; and $\omega_i \neq \omega_j, 0 \leq i < j \leq 2n-1$. Conjecture (but do not try to prove) what is the value of the limit

$$\lim_{n \to \infty} \tilde{K}_n^{1/2n}.$$

3-5. You are given an assignment to write a program which generates simple random walk paths in two dimensions of length 50. The program should use a random number generator and should produce walks uniformly distributed on the set of all simple random walk paths. In other words, when the program is run, any 50 step simple random walk path should be equally likely to be generated. How do you write such a program?

3-6. You are given an assignment to write a program which generates SAWs in two dimensions of length 50. The program should use a random number generator and should produce walks uniformly distributed on Ω_{50}. In other words, when the program is run, any 50 step SAW should be equally likely to be generated. How do you write such a program?

5-1. We defined the convolution of two probability distributions p_1, p_2 on S_N by the formula

$$(p_1 * p_2)(\pi) = \sum_{\lambda \in S_N} p_1(\pi \circ \lambda^{-1}) p_2(\lambda).$$

Show that convolution is associative, i.e., for any three probability distributions p_1, p_2, p_3,

$$p_1 * (p_2 * p_3) = (p_1 * p_2) * p_3.$$

5-2. Choose a completely random shuffle on S_{52}. What is the expected number of cards that stay in the same position after the shuffle? What is the probability that there is at least one card that is in the same position after the shuffle? What about S_N where $N = 2, 3, 10, 5000$?

5-3. A random walk on the symmetric group is called *ergodic* if every permutation can be reached eventually by the random walk, i.e., if for every permutation π there exists some j such that

$$\mathbf{P}\{Z_j = \pi\} > 0.$$

Consider random walks using the random shuffles in Examples 1 through 5. Which of these produce ergodic random walks?

5-4. Consider the following algorithm to shuffle a deck of cards on the computer. Start with the normal ordering $1, \ldots, N$. Choose a number j_1 at random from $1, \ldots, N$ and then transpose the cards in the 1 and j_1 position. Then choose another number uniformly on $1, \ldots, N$, say j_2, and transpose the cards currently in positions

2 and j_2. Continue this procedure using $j_3, \ldots$, the last step being to choose a number j_N uniformly on $1, \ldots, N$ and transposing the cards currently in positions N and j_N. Does this algorithm give a completely random shuffle, i.e., are all permutations equally likely to come up from this?

5-5. Here is another algorithm to find a random shuffle of a 52 card deck. Assume that your random number generator gives numbers between 0 and 1 to six decimal places. Choose 52 random numbers $x_1, \ldots, x_{52}$. Assume they are all different (if there is duplication, throw the numbers out and try again). Put these numbers in increasing order, say

$$x_{\sigma(1)} < x_{\sigma(2)} < \cdots < x_{\sigma(52)}.$$

Then output the permutation

$$(\sigma(1) \ \sigma(2) \ \cdots \ \sigma(52)).$$

Does this algorithm produce a completely random shuffle? If so, do you think it is better or worse than the algorithm discussed in Example 4?

5-6. Show that if you use a shuffle a deck of 52 cards four times using riffle shuffles that there are permutations π which cannot possibly occur. How many riffle shuffles are needed in order to have some chance of going from the natural order $1, 2, \ldots, 52$ to the reverse order $52, 51, \ldots, 1$?

6-1. After a single riffle shuffle (a 2-shuffle), what is the probability that the first card is kept in the same position? How about for an a-shuffle? Assume $N = 52$.

6-2. Compute D_1 exactly for $N = 52$.

6-3. Suppose one did a random walk using random transpositions as described in Lecture 5. Make a guess as to how many random transpositions are needed to make a deck close to completely random.

7-1. Let G be a finite group, and suppose p is a probability measure on G. Random walk on G (with increment distribution p) is the sequence of random variables

$$Z_n = Z_0 Y_1 Y_2 \cdots Y_n,$$

where Z_0 is some initial state, $Y_1, Y_2, \ldots$ are independent random variables with distribution given by p, and the multiplication is the group multiplication. Let

$$A = \{x \in G : p(x) > 0\}.$$

Suppose that A contains the identity, e, and that A generates the group (a subset of a group generates the group if every element of the group can be written as a product of a finite number of elements of the subset). What is

$$\lim_{n \to \infty} \mathbf{P}\{Z_n = e\}?$$

7-2. Show that any irreducible, aperiodic Markov chain on a set of two elements is reversible with respect to its invariant probability. Give an example of a three-state irreducible, aperiodic Markov chain that is not reversible.

8-1. Show that the pivot algorithm is irreducible. Give an example to show that if the only tranformations were rotations (i.e., no reflections allowed), then the chain would not be irreducible.

8-2. Let $G = (V, E)$ be a finite, simple connected graph, i.e., V is a collection of vertices and E is a collection of edges (unordered pairs of vertices) so that each edge connects two different vertices, any two vertices have at most one edge connecting them, and all vertices can be reached by a connected path of edges. Let f be a positive function on V. Give an example of a Markov chain such that: $P(x, y) = 0$ if $x \neq y$ and x is not adjacent to y (x and y are adjacent if there is an edge connecting them), and such that the invariant probability is

$$m(x) = \frac{f(x)}{\sum_{y \in S} f(y)}.$$

8-3. Consider a random walk on the symmetric group S_N induced by a probability measure p, i.e., the Markov chain with transition matrix

$$P(\pi, \lambda) = p(\pi^{-1}\lambda).$$

Assume that for all π,

$$p(\pi) = p(\pi^{-1}),$$

and hence P is a symmetric matrix. Show that for all integers n and all permutations π, λ,

$$P^{2n}(\pi, \pi) \geq P^{2n}(\pi, \lambda).$$

Show by example that this is not necessarily true for P^{2n+1}. (Hint: You may start by showing for any finite sequence of nonnegative numbers $a_1, \ldots, a_k$ and any permutation σ of $\{1, \ldots, k\}$,

$$\sum_{i=1}^{k} a_i^2 \geq \sum_{i=1}^{k} a_i a_{\sigma(i)}.$$

Then use the formula

$$P^{2n}(\pi, \pi) = \sum_{\sigma \in S_N} P^n(\sigma) P^n(\sigma).$$

and a similar formula for $P^{2n}(\pi, \lambda)$.)

8-4. Consider random walk on the complete graph on $N \geq 3$ vertices. This is the Markov chain on the state space $\{1, \ldots, N\}$ with transition matrix

$$P(x, y) = \frac{1}{N-1}, \quad x \neq y,$$

and $P(x, x) = 0$. Find all the eigenvalues and a complete set of eigenvectors for the matrix P. In particular, what is λ_2, the second largest eigenvalue?

8-5. Do the same thing for the following Markov chain: Let S be N-tuples $\mathbf{x} = (x_1, \ldots, x_N)$ with $x_i = 0$ or 1. Note that S contains 2^N elements. Let P be the transition matrix

$$P(\mathbf{x}, \mathbf{y}) = \frac{1}{N+1}$$

if $\mathbf{x}$ and $\mathbf{y}$ differ by at most one entry, and $P(\mathbf{x},\mathbf{y}) = 0$ if $\mathbf{x}$ and $\mathbf{y}$ differ in more than one entry.

8-6. Do the same thing for random walk on a circle: Let $N = 2k+1$ be an odd integer, let $S = \{0, 1, \ldots, N-1\}$, and let

$$P(i,j) = \tfrac{1}{2} \quad \text{if } |i-j| = 1,$$
$$P(0, N-1) = P(N-1, 0) = \tfrac{1}{2},$$

and $P(i,j) = 0$, otherwise.

11-1. Consider the following game: A die is rolled four times. If a '2' comes up at least once in those four rolls, one wins the game. What is the probability of winning the game? Do a simulation by playing the game 5000 times and seeing how many times you win.

11-2. Play the following game: A pair of dice are rolled until a sum of 5 or 7 comes up. If a 5 comes up first, you win; otherwise, you lose. Play this game 1000 times and see how many times you win.

11-3. Fit the following data to a function of the form (11.1).

x_i	y_i
5000	56.4
10000	79.8
15000	97.7
20000	112.2
25000	124.8
30000	135.5
35000	149.3
40000	159.8
45000	169.2
50000	178.4 .

11-4. How big is $\ln \ln x$ when x is the number of atoms in the universe? Does $\ln \ln x$ *really* go to infinity as x goes to infinity?

11-5. Consider the following game. You flip a fair coin until you get your first head. If it takes n flips to get your first head, then you win $\$2^n$. For instance, if you tossed a tail, then another tail, and then a head, you would win $\$2^3 = \$8$. Now we wish to simulate this game. Play the game 50,000 times (or more) and record the average winnings after 100, 1000, 5000, 10,000, and 50,000 games. Does it appear that the average winnings are converging to some number? How much would you be willing to pay to play this game once? How much would you be willing to pay to play this game for as long as you wished? Try to find a formula that will estimate your average winnings after n games.

12-1 You are fishing on a Saturday morning. The time between catching each fish is an exponentially distributed random variable with parameter $\lambda = .8$. You will be fishing for 4 hours. Simulate the number of fish caught by sampling from an exponential distribution. (Here is how the algorithm goes: Take a sample from an exponential distribution. If its value is greater than 4, you catch no fish; if the value is $t < 4$, you have caught one fish, and have time $4 - t$ to catch more fish. Continue until four hours are up.) Do the simulation for at least 1000 Saturdays, and estimate the probability mass function of the number of fish caught (i.e., probability of catching no fish, 1 fish, 2 fish, etc.)

Compare your results to the probability mass function of a Poisson random variable with parameter $\lambda = 3.2$. (Why would we want to compare them?) The probability mass function for the Poisson r.v. is

$$p(n) = \frac{e^{-\lambda}\lambda^n}{n!}$$

12-2. Write a program which generates random samples from the normal distribution with $\mu = 0$ and $\sigma^2 = 1$ (i.e., from $N(0,1)$). Generate at least 2000 such samples. For each of the intervals $(-\infty, -1.5)$, $(-1.5, -1)$, $(-1, -.5)$, $(-.5, 0)$, $(0, .5)$, $(.5, 1)$, $(1, 1.5)$, $(1.5, \infty)$, record how many of the random samples fall in these intervals, and thus estimate the probability that a random sample will lie in each interval. Compare with the exact probabilities.

12-3. Suppose that are 14 people in a family that wish to exchange gifts. Each person will give one gift to one other person. Each family member's name is put in a hat, and then each person picks out one name (at random) from the hat. If person A picks out person B's name, then A buys a gift for B. Write a program which will generate a random permutation and proceed to simulate the picking of the names *at random*. Run this simulation at least 2000 times and each time keep track of how many people get to buy a gift for themselves. Use this to estimate the probability mass function for the random variable

$$X = \# \text{ of people that pick their own name.}$$

13-1. Suppose that the price of a certain stock can be modeled using geometric Brownian motion with a drift rate of 13% per annum and a volatility rate of 27% per annum. Using the first method for simulating the path of the stock price that we discussed in Lecture 13 (i.e., using equation (13.4)) with $\Delta t = \frac{1}{365}$ (i.e., one day) to generate 1,000 sample paths for the stock price over the next four months, given that the current stock price is \$55. Graph at least one of the paths. Record the final price (i.e., the price after four months) for each of the paths and check that the average is approximately equal to $55e^{(.13)4/12} \approx 57.44$.

13-2. Repeat question 13-1 using the second method that we discussed in Lecture 13 (i.e., using equation (13.5)).

13-3. Suppose that a stock price has an expected return of 15% per annum and a volatility of 29% per annum. If the stock price at the end of today is \$60, approximate the following using equation (13.3):

(a) the expected stock price at the end of tomorrow,

(b) the standard deviation of the stock price at the end of tomorrow, and

(c) a 95% confidence interval for the stock price at the end of tomorrow.

13-4. Suppose that stock A and stock B both follow geometric Brownian motion and that changes in any short interval of time are uncorrelated with each other. Does the value of a portfolio consisting of one of stock A and one of stock B follow geometric Brownian motion?

Bibliography

[BD] Bayer, Dave and Diaconis, Persi, Trailing t
lair, Advances in Applied Mathematics **8** (19

[CW] Calkin, Neil and Wilf, Herbert, The numbe
a grid graph, SIAM J. Discrete Mathematics

[DS] Doyle, Peter and Snell, Laurie, *Random W*
works, Carus Mathematical Monograph 22, M
of America (1984).

[H] Hull, John, *Options, Futures, and Other Deri*
Hall (1996).

[L1] Lawler, Gregory, Random walks: simple an
(1995), 55–74.

[L2] Lawler, Gregory, *Intersections of Random W*
(1991).

[L3] Lawler, Gregory, *Introduction to Stochastic*
Hall (1995).

[MS] Madras, Neal and Slade, Gordon, *Th*
Birkhäuser-Boston (1993).

[M] Mann, Brad, How many times should you sh
Snell (1995), 261–289.

[P] Pemantle, Robin, Uniform spanning trees, in

[Si] Sinclair, Alistair, *Algorithms for Random C*
Birkhäuser-Boston (1993).

[S] Snell, Laurie, ed., *Topics in Contemporary P*
cations, CRC Press (1995).

Bibliography

[BD] Bayer, Dave and Diaconis, Persi, Trailing the dovetail shuffle to its lair, Advances in Applied Mathematics **8** (1987), 69–97.

[CW] Calkin, Neil and Wilf, Herbert, The number of independent sets in a grid graph, SIAM J. Discrete Mathematics **11** (1998), 54–60.

[DS] Doyle, Peter and Snell, Laurie, *Random Walks and Electrical Networks*, Carus Mathematical Monograph 22, Mathematical Association of America (1984).

[H] Hull, John, *Options, Futures, and Other Derivatives*, 3rd. ed., Prentice Hall (1996).

[L1] Lawler, Gregory, Random walks: simple and self-avoiding, in Snell (1995), 55–74.

[L2] Lawler, Gregory, *Intersections of Random Walks*, Birkhäuser-Boston (1991).

[L3] Lawler, Gregory, *Introduction to Stochastic Processes*, Chapman and Hall (1995).

[MS] Madras, Neal and Slade, Gordon, *The Self-Avoiding Walk*, Birkhäuser-Boston (1993).

[M] Mann, Brad, How many times should you shuffle a deck of cards?, in Snell (1995), 261–289.

[P] Pemantle, Robin, Uniform spanning trees, in Snell (1995), 1–54.

[Si] Sinclair, Alistair, *Algorithms for Random Generation & Counting*, Birkhäuser-Boston (1993).

[S] Snell, Laurie, ed., *Topics in Contemporary Probability and its Applications*, CRC Press (1995).

(c) a 95% confidence interval for the stock price at the end of tomorrow.

13-4. Suppose that stock A and stock B both follow geometric Brownian motion and that changes in any short interval of time are uncorrelated with each other. Does the value of a portfolio consisting of one of stock A and one of stock B follow geometric Brownian motion?